CREATIVE CAREERS IN HOLLYWOOD

CREATIVE CAREERS IN HOLLYWOOD

LAURIE SCHEER

ALLWORTH PRESS
NEW YORK

07 06 05 04 03 02   5 4 3 2 1

Published by Allworth Press
An imprint of Allworth Communications, Inc.
10 East 23rd Street, New York, NY 10010

Cover design by Mary Ann Smith, New York, NY
Page composition/typography by Integras Software Services Pvt Ltd., Pondicherry, India
ISBN: 1-58115-243-4
Library of Congress Cataloging-in-Publication Data
Scheer, Laurie.
Creative careers in Hollywood / Laurie Scheer.
p. cm.
Includes bibliographical references and index.
ISBN 1-58115-243-4
1. Motion pictures--Vocational guidance--United States. 2. Television--Vocational guidance--United States. I. Title.
PN1995.9.P75 S34 2002
791.43'02'93--dc21
2002011071

Printed in Canada

SPECIAL THANKS TO:

Anais, Ayn, and Hortense, who paved the way,
Daisy and Sir, who were the way,
and to Professor James Tiedge, who taught me the way.

To DePaul University and Dr. Charles Strain and Dr. Kate Kane.
To the University of Chicago for letting me teach this material for the first time in the spring of 1997 and to all those who took the course. To Northwestern University for giving me "my kids." To Dr. Deena Weinstein, warmly, for believing in me. To Gahl who never stopped asking how the book was going. To Cindy Bell who gave this Midwesterner her first job in the industry. To Kelley Rarey, who is no longer from Ohio. To Allworth Press for believing in the project. To the Margaret Herrick Library, Center for Motion Picture Study, and to Eddie Brandt's Saturday Matinee Video. And to Mom and Dad for those typing lessons at age twelve.

To every kid who has ever wanted to work in the movies,
this book is for you.

The journey-to-life of this book began while I was in graduate school at the University of Chicago. Part of the agreement to complete a master's degree in liberal arts was for a graduate student to teach the material she was researching for her thesis. The university generously granted me the opportunity to impose my findings, musings, bits and pieces of media, and video bytes on whoever entered the classroom that spring. The following were the gracious individuals who participated in the adventure. I thank them for helping to give life to this information.

"HOLLYWOOD ON HOLLYWOOD"
Gleacher Center—University of Chicago
Chicago, Illinois   March 22, 1997

Anne Allison

Diane Dawson

Denise Doctor

Tom Foley

Jennifer Garson

Joanna Garson

Beau Grabner

Judy Lapinsohn

Jeffrey Meyer

Ellen Nordberg

Jacquey Rosati

Susan Roupp

Cheryl Lyn Rybka

Ruth Schoenbeck

Robvert Schrade

Douglas Stevenson

Samantha Weaver

CONTENTS

INTRODUCTION

If there's one thing I hate, it's the movies. Don't even mention them to me.
The goddamn movies. They can ruin you. I'm not kidding.
—Holden Caufield, Catcher in the Rye

The movies can ruin you; there is no doubt. It is probably true that most people working in Hollywood have been ruined by the movies—and that's a good thing. Many American children in average All-American households are introduced to movies and media early on in life and become addicted by adolescence. What would life be like without movies to watch? Without television, cable, and the Internet? We are all influenced in some way by the presence of media in America; however, some members of the general population take the fixation to the next step and attend colleges and universities to obtain degrees in Communication, Media Studies, or Film and Television tracks. Upon graduation, they are determined to go to the coasts and media hubs of America to explore their future behind the desks and cameras of the entertainment industry. They are not alone. They have been preceded by many and will be followed by many more. The working hotbed of the entertainment industry is Los Angeles. If you want to be in the movies, this is the place to be, or at least a place to begin.

THIS IS LOS ANGELES...

Hollywood is a city, an industry, and a state of mind. It is an immense background scenario and a perpetual pan shot that recycles and reinvents itself continuously. Los Angeles was just an arid basin at the beginning of the twentieth century, populated with nothing but orange-tree groves. Those sophisticated folks up north in San Francisco often referred to their southern neighbor as a "cow town." Between 1885 and 1915, Los Angeles saw the arrival of the Santa Fe Railway, the real-estate frenzy of 1886–88, discovery of oil in her earth, and the first national ad campaign of the region's citrus growers. Finally, Los Angeles welcomed the arrival of early film producers, who found the climate just right for the production of a new art form, which evolved into a very lucrative industry.

A Los Angeles suburb known as Hollywood was the birthplace of moviemaking and became the heart of this booming new industry. The first "Big Five" studios popped up, and the studio system got underway. Hundreds and thousands of newcomers, arrivals from other states and countries, descended upon Southern California to work in this new industry. Young people with skills in carpentry and painting arrived, alongside those seeking fame and fortune by virtue of their looks. There is a common saying: "Like everyone from Southern California, he's not from here." They came to work in the movies. They were the pioneers, the early settlers, those who blazed the trail for the others to follow. They made up the rules along the way. Today people continue to arrive daily to attain that dream and work in the entertainment industry. What was started over a hundred years ago continues to thrive because of all the people ruined by the movies.

MOVIES AREN'T THE ONLY INFLUENCE

As the decades passed, the early pioneers gave way to the journeymen of the thirties, forties, and fifties, who themselves had been raised on watching movies. The fifties, sixties, and seventies had exposed post–World War II arrivals not only to moviemaking but also to the wonders of television. Finally, by the eighties and onward, the worlds of cable and the Internet became reality. American entertainment, mostly headquartered in

Hollywood, where the dreams are constructed, has become a major global business.

The Dream Factory—that's what the entertainment industry really is for the vast majority of people working in movies, in a creative environment of mythmaking. Not every laborer in Hollywood is aware of his part in molding the collective consciousness, but every individual there plays a part in constructing the entertainment, and ultimately the myth. Hollywood exists first and foremost within our collective consciousness. And it is because of the hundreds of thousands of individuals who continue to want to work in Hollywood and pursue their lives' dreams that the myth is perpetuated. All of these individuals are both *in* the movies and *of* the movies.

MYTHMAKING FOR THE COLLECTIVE CONSCIOUSNESS

Why all this talk about working in Los Angeles, making myths, and being part of the collective consciousness? Because our modern-day movies are tangible evidence of our modern-day myths. The movies are written, produced, and directed by individuals, who are part of our society; they are not manufactured by the gods above, but rather, by artists who are just like you and me. So, in order to learn about life in Hollywood, it might be interesting to explore the movies *about* working in the movies, which have been made by people who work in the movies. The lifestyle and adventures of working in film will be revealed within the very medium the industry produces.

And so this book was born. Out of the history of film during this past century, I have chosen movies featuring the basic jobs of the entertainment industry—Actor, Agent-Manager, Assistant, D-girl, Director, Press, Producer, Production and Crew, Studio Executive, and Writer. Each illustrates one or more of the above-named jobs. The duties, the trials and tribulations, the highs and lows, and the rewards of each job are explored as the positions are scrutinized. The result is a patchwork account of the creative career choices one has access to when working in Hollywood.

Analyzing movies made about working in the movies is challenging and fascinating at the same time. It has often been said that in order to go forward, it is best to look backward so the same mistakes aren't made. This book will provide a place to begin to understand the triumphs and defeats of all of the on-screen workers in the industry.

1950, 1976, AND THE LATE 1990S

The concept of Hollywood turning the cameras and attention upon itself is not new, and many fine books, articles, and research material are available on this topic. However, the concept of using these same films as a way to learn about the jobs within the entertainment industry is new. There have been three important movements in American cinema during the past fifty years wherein filmmakers have provided key behind-the-scenes films.

The first was during the fifties, when there was a surge of "big" films about Hollywood. The reason for this trend was the onslaught of the television industry. The free flow of entertainment right into the public's living room threatened the studios. Moviemakers were forced to go behind the scenes in the fifties to continue to hold on to the public's interest. *Sunset Boulevard, The Bad and the Beautiful, The Goddess, The Star, All About Eve, A Star Is Born* (1954), *Beloved Infidel,* and *The Big Knife* were all produced during this time. It was also during this time that the first and only book reflecting an anthropologist's study of Hollywood, *Hollywood: The Dream Factory* by anthropologist Hortense Powdermaker, was written. This book, which could be considered an ancient predecessor to some of today's popular behind-the-scenes projects (both written and filmed), is a unique sociological study. Ms. Powdermaker took a year to live and work among the people who dedicated their lives to working in Hollywood, and recorded her observations for posterity. Most of Ms. Powdermaker's findings remain true today, more than half a century later.

The second resurgence of interest in films about the entertainment industry took place in the seventies, when a flock of movies about Hollywood and its past appeared. During the mid-seventies, audiences were treated to *The Wild Party, Day of the Locust, Hearts of the West, Inserts,* and *The*

Last Tycoon. None of these films did boffo box office. It was a strange trend for the time when most of the moviegoing public was engrossed in the phenomenon of *Star Wars*. Nonetheless, Hollywood is narcissistic, and there were a number of filmmakers who came up with the same idea — to look at themselves and at the industry's history. Most members of the current generation of filmmakers were just getting their start in the mid-seventies. Their interest in the business led them to create movies about the stories and the stars of cinema history that had influenced their decisions to work in the industry. The second generation of Hollywood-working souls wanted to celebrate their work and bring it to the big screen. Only a fraction of the public appreciated their vision.

Finally, within the last decade of the twentieth century, yet another wave of filmmakers decided to turn the cameras onto themselves and again produce important films about their generation and their industry. Among these films are *The Big Picture, Living in Oblivion, Swimming with Sharks, Barton Fink*, and *Swingers*. Most of these films were produced on a low budget. In addition, the popular *The Player*, plus *The Truman Show, Notting Hill*, and *Bowfinger* appeared at the box office. This time, the audience was more supportive, and these films quickly received audience and critical acclaim. Some have become cult films — mostly due to film buffs, insiders who have a love-hate relationship with the industry, and fans who admire the motion-picture–show-business machine.

There is every reason to believe that this trend will continue. Those of you reading this book, who do decide to go on to careers in film, may well perpetuate the trend and go behind the scenes to tell us of your life in the entertainment industry.

Almost everyone knows two or three films about working in Hollywood. Films such as *Sunset Boulevard, The Player*, and, if you ask those of a younger age, *Swingers, Living in Oblivion*, and *Swimming with Sharks* are just a few that come to mind. There are, however, hundreds of films about working in the entertainment industry. What follows is an analysis of those that best illustrate each of the prominent positions. It is true that Hollywood does influence the collective consciousness of the world. Each of these movies has undoubtedly influenced generations of individuals to choose work in the entertainment industry.

LIFE DOES INFLUENCE ART

As this book's deadline was being met, America endured the tragic attacks that occurred on September 11, 2001. Within the days and weeks following the attacks, an arts-and-entertainment task force was summoned by the White House and government intelligence specialists to help them brainstorm about possible future attacks from the enemy. As Allison Hope Wiener and Daniel Fierman wrote in an *Entertainment Weekly* article, "Since 1999, the Institute for Creative Technologies (ICT) at the University of Southern California, has been operating as a brain trust for the U.S. Army, working in conjunction with academics and unpaid volunteers from the entertainment industry, including David Fincher and Spike Jonze as well as special-effects gurus like Ron Cobb. The institute, headed by Richard Lindheim, a former executive at Universal and Paramount, helps create virtual-reality-training experience for soldiers. (The ICT is operated by USC under a five-year, $45 million contract with the Army.) While terrorism has long been on the ICT's agenda, industryites have met since September 11 to intensify their brainstorming about possible attack scenarios."[1]

Surreal? No, not at all. Those who create movies include in them circumstances from real life. Yes, reality is at times masked by elements of storytelling and embellished for the sake of drama, but often movies become our contemporary icons and are soon reflected upon as truth. It is no surprise that the government has turned to creatives for answers, for they have been writing violent, terrorist-filled movies for the last few decades. The fact that the government turned to entertainment industry workers during our national crisis shows that we believe that the movies have something truthful and perhaps even visionary to say about contemporary life.

The movies we are about to discuss do, in fact, hold many truths about work in the entertainment industry. Many hard-working writers, directors,

---

[1] Wiener, Allison Hope and Fierman, Daniel, "Marching Orders," Entertainment Weekly, October 19, 2001.

crew people, and actors alike have toiled over these films to bring you, the movie buff and student of the entertainment industry alike, a realistic and reflective look at what it's like to work in Hollywood.

TOUCHSTONE THOUGHTS

This book uses the medium of film to teach the next generation of filmmakers and their audiences about films of all genres—and the behind-the-scenes jobs that add up to the monumental task of moviemaking. Each chapter is a celebration of a position, giving examples of films that shine a spotlight on it, then breaking these films down according to the decade they reflect, and finally observing and comparing the differing ways in which the filmmakers chose to present their material. From the discussion of each film throughout respective chapter, the reader should come away with a pretty good idea of what the job entails.

With each observation, I, your humble author, am merely pointing the way. As an individual who has had many experiences in the Hollywood arena, I am only able to share, to lecture, to teach, and to open the door to this world of celluloid movie work. You, dear reader, will pass through that door by yourself.

This book is for dreamers, risk-takers, and others who laugh in the face of normalcy. Anybody with imagination and no fear is welcome to join the party. There will continue to be an infinite number of newcomers and wannabes arriving in Los Angeles and all of the media hubs in the near and far futures. This book is for you. And please remember, "Your dream isn't big enough." So, start reinventing, readjusting, and reacting to a whole new world of opportunities. Be a pioneer. Blaze your own trail. Start now.

HOW TO USE THIS BOOK

Each chapter of this book presents a particular Hollywood profession, and then discusses some of the films that focus on that profession. The films are listed by the decade they represent, from the beginning of film history until 1999. Each chapter will then examine the following:

- The duties and function of the job
- The various levels of the job
- A summary and brief history of the job
- A discussion of the movies that feature the job being explored
- Interviews and observations about endearing quirks peculiar to each job

Also, at the beginning of each chapter you'll find a Creative Careers in Hollywood Status chart, which will help you place the position in the Hollywood Food Chain.

HOLLYWOOD FOOD CHAIN (HFC)

"Shut up, listen, and learn!" is Buddy Ackerman's mantra in the 1994 indie flick *Swimming with Sharks.* Kevin Spacey plays a hard-ass Hollywood executive who doesn't seem to have a soul. *Sharks* is an example of a quintessential working-in-Hollywood movie. All those who have ever thought of being an assistant or working their way up the Hollywood Food Chain need to see this somewhat dark but not completely fictitious account of an employer-slave relationship. And while one need not shut up, necessarily, one may, in perusing these pages, listen, and one may learn something about what it's like to work behind the scenes in this dream industry, this mythmaking machine.

To begin to understand Hollywood, one should keep a copy of the following Hollywood Food Chain on hand for reference while reading this book, for what follows is a very valuable chart that could be referred to as the "Chutes and Ladders for the Hollywood Set." At various times in your career, you could find yourself at the polar opposites of this food chain. It may be best to study it now, so all is familiar when your turn arrives to "work the room" or "climb the ladder of success" or "bottom out due to a bad business deal."

Coinciding with the Hollywood Food Chain are the agents who run it, a.k.a. those who negotiate the deals, those who have the power (see the Agent-Manager chapter). Please note that there are agents that specialize in and within each of the categories. Here's an example. One agent may

THE HOLLYWOOD FOOD CHAIN

site	who?	budget*
1. Feature Films (studio)	The Big 8**	$10–25 million to $170 million plus
2. Feature Films (indie)	New Line, Miramax, Artisan	Under $20,000 to $20 million
3. Cable (premium channels)	HBO, Showtime, TMC, etc.	$2–50 million
4. Cable (basic)	MTV, ESPN, Lifetime, CNN	Series: $25,000 plus per half-hour Specials: $100,000 plus per hour Movies: $2 million
5. Networks (broadcast)	ABC, NBC, CBS, FOX	Series: hundreds of thousands Movies: $2–10 million Miniseries: $5–20 million
6. Weblets (broadcast)	UPN, WB	Series: $50,000 plus per half hour
7. Local Channels	KTLA, KTTV, KABC-LA	Budget reflects local ad sales
8. Cable Channels	Start-ups	Little to no budget, use reruns
9. PBS	Varies locally	You pay them to be on air

*Budgets are grand approximates and vary from company to company, year to year

**The Big 8 Studios are: Disney, Dreamworks, MGM/UA, Sony/Columbia, Twentieth Century Fox, Universal, Viacom/Paramount, and Warner Bros.

represent only feature film (mostly studio) directors and would not think of getting involved in a deal other than a potential feature film project for his client. Another agent only reps television sitcom writers and would not be heard negotiating anything other than sitcom possibilities for clients. Hence, if one is planning to work as an actor, writer, director, or sometimes a producer, one needs to find the corresponding agent who works in the right arena. Additionally, a first-time writer, producer, or director will often be pigeonholed into one level of the HFC due to the success he found in the arena he was first discovered in within the industry. Career mobility will increase when that initial success spreads across the various levels of the

HFC, such as a popular feature film that becomes an even more popular television show, which in turn produces successful merchandising. Equally, should a first-time project open to a small box-office return or low ratings, a career could be destroyed at the gate. There are, however, new opportunities for adventure and for expansion of one's career in the number of outlets available to a film after its traditional release via a studio or network. Consider, for instance, such outlets as:

- Domestic release
- Foreign release
- Pay-per-view
- Rental (home video and DVD)
- Cable premiere
- Cable basic run
- Network
- Syndication

Within each studio, cable network, and network, there is another layer of the Hollywood Food Chain, which consists of the above-listed departments. In other words, if one works in the Foreign Release or Home Video department of the studio, it is not as grand as working for the VP of Production for the studio. Additionally, working in the mail room at an agency or studio is far more illustrious than working as an assistant to a mid-level manager at a television network, and certainly more desirable than working in the mail room at a local channel or PBS station. So one should choose wisely when one begins the Hollywood Chutes and Ladders game. The following chapters will arm you with valuable information. You'll see how others have performed these jobs in the movies made about making movies, and you'll learn how to climb high on the Hollywood Food Chain. Don't be a bottom-feeder!

DON'T FORGET TO HAVE FUN

And finally, don't forget to have fun. Use this book as a tipping point, a place to begin your research into what the industry is all about, whether you intend to work within the walls of a studio or you are just plain curious about what goes on behind the scenes. Joseph Campbell, the great lecturer and writer who deconstructed so many of our myths, coined the phrase "Follow your bliss." If your journey is to follow the paths of the individuals discussed within these pages and seen on the silver screen, then so be it. Follow that bliss. Always and forever . . . and have fun while you're doing it. If you stop having fun, then stop doing it.

CHAPTER 1

ACTOR

STATUS

DURABILITY: Shredder.*

LENGTH OF STAY: That's a tough call. You could be an extra, a day-player, or part of the lead cast—it varies. The real question is how many years of your life do you devote to the art of auditioning?

FOOD-CHAIN VALUE: High if you make $20 million, low if you are a day-player.

UPWARD MOBILITY: Not a lot.

DESIRABILITY FACTOR: High, especially among new arrivals from Ohio.

VACATION: None, actually—you are on vacation all the time, and you usually work one week out of the year.

SALARY: Tap water to bottled water from France.

HOW EASY IT IS TO GET THIS JOB: On a scale of 1 to 10 (1 being the easiest), 10.

PREREQUISITES: To be really, really good looking. To have charisma. To be able to charm the pants off of any casting agent. To be intellectually challenged. To find rich old producers to sleep with.

We didn't need dialogue. We had faces.
—*Norma Desmond,* Sunset Boulevard

Actor, artist, performer, entertainer, thespian, and star—just a few names for those bold souls who have taken on the job of relaying the emotion and action of any given script onto the silver screen. Only a percentage of those members of the Screen Actors Guild work on a consistent and steady basis—a very small percentage. Yet, year after year, the number of wannabe-actors who begin their journeys toward fame grows. For actors, as the myth goes, are nearly godlike and have more, mean more, and represent more than any ordinary mortal and therefore deserve to be celebrated on the silver screen and elevated to a level of superhuman.

This chapter illuminates the occupation of acting. It follows the trials, tribulations, and joys of actors by taking a look at the films about working in Hollywood as an actor. As we analyze each decade, the growth and changes of this position, which have taken place over the last century, will become evident. The job of an actor is exciting, adventurous, oftentimes difficult, and, overall, extremely important.

The definition of actor is "one who acts." It is simply a matter of the amount of screen time an actor gets that varies, and this variation adds

There are some creative careers within Hollywood that are going to be keepers, and there are some creative careers that are going to be shredders. And there are creative careers that are confusing: People may think they are keepers, when in reality they're shredders. (Keepers are just that—jobs that are highly desirable and, once obtained, must be held on to at all costs; shredders, on the other hand, are jobs that should be ripped to shreds or ended as soon as the usual length of stay has been completed.)

degrees and definition to the job. An extra, or day-player, is just that, an actor hired for a scene or a day only. A supporting actor is a character within a movie supporting the lead or lead cast, and a leading actor is the main star. Here, then, are the movies that focus on the profession of acting.

THE TWENTIES

Three movies representing the Roaring Twenties give us a taste of what Hollywood was like at the time of its infancy. Los Angeles and her suburb of Hollywood were one-horse towns. With the backdrop of orange and lemon groves behind them, movie people moved in and built their studios. Hollywood in its immaturity was a land filled with pioneering souls who had ventured across the country and from throughout the world to explore this new industry.

The Extra Girl (Associated Exhibitors, 1923)

The Extra Girl, featuring comedienne Mabel Normand, is the story of Sue Graham, a young girl who wins her Illinois hometown beauty contest and, as a result, earns a trip to 1923 Hollywood. After a short time, she fails at finding steady work in front of the camera and becomes an assistant in the studio wardrobe department. She works sporadically as an extra girl, but that's the only pseudofame she can find. She is given brief walk-on and background parts only. All of her work is nonspeaking and behind-the-scenes, far below her superstar aspirations.

Her escapades on the lot are funny for their day. Sue Graham is essentially a "female Merton" (we'll meet Merton, a famous actor-wannabe character, in the next few pages), and her wide-eyed innocence reflects the twenties' mindset. She fails at achieving leading-lady status, but finds comfort in writing home to her Illinois family of the glories of Southern California. When miniscule extra work seems to be the only kind of acting work she can muster up, Sue eventually gives up and returns home, leaving her dream behind. Leaving Hollywood so soon would not be recommended these days—at least not until all levels of the Hollywood Food Chain were explored.

Show People (MGM, 1928)

King Vidor's Hollywood expose featuring lead character Peggy Pepper (Marion Davies) was intended as a gentle satire on the career of Gloria Swanson. Ms. Swanson found herself in an emerging industry in the beginning of her career, an industry that focused heavily on comedy, and comedy was not really what she had wanted to do. By the end of Swanson's on-screen life, she had bridged the gap between the silent movies and the talkies—one of the few stars to survive the transition. Swanson went to the extreme opposite of her early comedic persona and became a serious on-screen thespian, which many in the industry found hard to accept. She also, it has been said, took on an attitude that left her slapstick pals behind, leaving room for a lesson in humility.

Peggy Pepper is the character The Extra Girl would have evolved into if she hadn't packed up and returned home to Lincoln's birthplace. Southern belle Peggy is a wide-eyed innocent, struck by the bright lights of the movie set. She can only find employment as Billy Boone's (William Haines) sidekick. As Boone falls in love with her, he teaches her how to be a serious actress. Peggy finds success as a dramatic actress, leaving her slapstick scenes in the dust as her ego grows to new heights. Boone brings her back to earth, reminding her of her roots and she soon learns her lesson.

Director-producer King Vidor's *Show People* is a movie that defends the Hollywood world it portrays. Irving Thalberg produced this film, although his work is uncredited. *Show People* is an important film to see to understand the silent-film era, and its title is ever so appropriate. This newly created town, fast-developing the habit of producing dreams, came under criticism for being a hotbed of immorality. Show business produces show people who are not unlike carnival and circus acts; they just have different types of lions and tigers and bears to work with. This insider's view is one of the earliest reflections of working in Hollywood, and it remains one of the best. The movie also provided Ms. Davies with an opportunity to explore her acting range—and it didn't hurt MGM that her every move was hyped by her admirer William Randolph Hearst in all of his many newspapers.

The Wild Party (American International, 1975)

This comedy-drama-musical tells the tale of a wild party that took place in 1929; a staged version of what some say reflects the notorious scandals of the same time period. Heavy with dialogue and garish, baroque sets (similar to *Inserts,* see chapter 5), this movie shows why in the twenties it didn't matter to the public if an actor stole money, made love to a member of his own sex, fathered a child out of wedlock, or killed himself. The industry was still too young and the publicity machine was not in place yet. Audiences saw actors in their silent films and pretty much accepted them as one-dimensional people. What was happening locally in Hollywood, however, was another story.

The Wild Party provides scene after scene of movie people acting out large, screen-sized fantasies. Some are bored, some are drunk, and some are practicing overall debauchery. The movie is peppered with graphic and erotic heterosexual and homosexual scenes. There are orgies going on everywhere with cocaine and cock galore. This movie takes place during a twenty-four-hour period and is narrated in rhyme based on the 1928 poem "The Wild Party" by Joseph Moncure March. Released in 1975 and featuring Raquel Welch, James Coco, Perry King, and David Dukes in the leading roles, *The Wild Party* was a box office dud.

The movie does reflect the energy of the people who worked in Hollywood at this time. After the long days on the set, the opulence and extravagance of Hollywood kicked in. These people knew how to party. The party scene never really went away; Hollywood people still like to celebrate.

THE THIRTIES

This decade features some of the richest films about acting. With *What Price Hollywood?* and its remake, *A Star Is Born,* this is the decade to watch for some of the best tips about how to prepare to be a movie star.

What Price Hollywood? (RKO-Pathé, 1932)

Mary Evans (Constance Bennett) is a waitress at the famous Brown Derby restaurant. She is also a struggling actress, so when she helps a very drunk

Maximillan Carey (Lowell Sherman) get home safely, she is thrilled to learn that he is a successful movie director. Their friendship grows—this is 1932, so it is strictly platonic—and Max helps Mary to stardom. This is the first of the many renditions of this story, the classic tale of the ordinary but talented young girl who gains fame while the seasoned older man who helped her get there dwindles into mediocrity and meets a tragic ending. It is the story of the chutes and ladders of Hollywood—as one climbs the ladder to fame, the other swiftly finds himself gliding down the ladder to obscurity. These eighty-eight black-and-white minutes are a joy to watch. The dialogue is snappy and pompous and the performances are sharp. *What Price Hollywood?* is a gem of a film that is mandatory viewing for anyone interested in working in the entertainment industry.

A Star Is Born (United Artists, 1937)

This Dorothy Parker script is nearly perfect in its attempt to give an answer to the question posed in its predecessor, *What Price Hollywood?* It is the big-screen rendition of the star-rising-to-fame tale, and it has the classic scenes of the young girl arriving in Hollywood to realize her dream of being a star—and paying the price for that opportunity.

In Hollywood, the metropolis of make-believe, Esther Blodgett (Janet Gaynor) arrives and immediately needs to find a place to stay. She answers an ad at the Oleander Arms—LARGE ROOMS, RUNNING WATER, NO COWBOYS. For $6 a week she rents a room and takes on a switchboard job, constantly checking the want ads and call sheets for an acting gig. "Chances are one in one hundred thousand, but maybe I'm the one . . . ," she tells herself, "I could make them notice me." And yes, Hollywood does notice her, through Norman Maine (Fredric March), a self-destructing matinee idol. As her star rises, once again, his star fades, but before that happens, they have a heck of a good time painting the town and living the high life. Esther becomes Mrs. Norman Maine and also renames herself Vicki Lester.

What is the lesson to be learned from these two films? Life for young hopefuls today is not that different from what it was for Esther. Seventy-plus years later, just about every aspiring actress does waitressing and lounge-singing gigs, hoping to be that one-in-one-hundred-thousand girl to be called back, to be a star. The film may be grainy and old, but the journey from ordinary civilian to glamorous star remains the same.

Going Hollywood (MGM, 1933)
Showgirl in Hollywood (First National, 1930)

In 1930 alone, one year after the stock market crash, no fewer than seventy musicals were filmed in Hollywood. Dispirited Depression audiences were hungry for an escape from the meanness of everyday life. The fantasy worlds and fairy-tale plots of Hollywood musicals provided a much-needed diversion. *Showgirl in Hollywood* is one version of this story.

Going Hollywood, is another musical variation on the *What Price Hollywood?–A Star Is Born* theme. This uncomplicated boy-meets-girl musical production features Marion Davies as Miss Sylvia Bruce, who lives to experience music, love, and life. She is tired of being a schoolteacher and abandons her career to find her idol, Bill Williams (Bing Crosby). Talk about lofty ambitions! Thus her dream begins, and a huge send-off takes place on the steps of Grand Central Station. Fast forward to her arrival in L.A., followed by a beautiful montage of famous Hollywood sites in full black-and-white glory.

Sylvia is clearly enamoured during her star search. She finds Bill through his frequent costar and jealous girlfriend Lili (Fifi D'Orsay). Lili takes Bill to Mexico to get away from Sylvia, but Sylvia cannot be held back and follows them relentlessly. She persuades Bill to return to Hollywood and finish a musical—he does so—without Lili, and now Sylvia is Lili's replacement in the movie. She realizes her dream. She gets to sing and dance with her idol in a true Hollywood "happily ever after" scenario.

Make Me a Star (Paramount, 1932)

The male point of view on becoming an actor in the thirties can be seen in *Make Me a Star.* This film tells the story of one Merton Gill (Stuart Erwin), a grocery-store clerk in a small town in Illinois. He studies acting through a correspondence course and is dedicated to his dream of becoming a movie star. When he loses his job at the store, he gets on a train and follows his dream. Upon arriving in Hollywood, he finds his way to the studio employment office and offers what little he knows about amateur acting. The office workers make fun of him. They can't believe he is for real . . . but he is.

Merton faces continual rejection, but then, through a twist of fate, he is cast to star in a Western satire. Merton plays his cowboy role straight, honored that he has been chosen for this role. Merton takes his acting very seriously, but audiences do not. He thinks he is a failure as an actor, but the casting directors realize that he is a comedic genius. At first, Merton is insulted that his dramatic potential has not been recognized, but ultimately he accepts the fact that he has found success as a comic—not a tragedian.

Merton of the Movies (MGM, 1947)

This 1947 production, a remake of the 1924 movie of the same name and of *Make Me a Star*, was a star vehicle for comic Red Skelton. Some baby boomers may remember Skelton as a television funnyman, yet in the days before the small screen, Skelton had a substantial movie career. In this film, he plays the Merton Gill character, but this time he is an usher from Illinois who has just arrived in Hollywood. He befriends an extra girl and gains access to a movie studio where he is cast for a comedic role that he feels he is not right for because he has always wanted to be a dramatic actor. He returns to the Midwest, having failed his dream. This movie is the embodiment of the Young Thing (in this case, a Boy) Wants Hollywood plot. Boy gets Hollywood dream; boy becomes disillusioned with Hollywood dream; boy returns home.

Many of these early movies about being a star feature naïve and innocent individuals who follow their dream only to face disappointment and rejection. In this case, Merton is not tough enough to endure being cast outside of how he sees himself as an actor. Industry people have been "typecasting" actors for as long as there has been a movie industry. It is a lucky actor who fits easily into being a "marketable type." This is a business that thrives on survival of the fittest—it was like that in the twenties and continues a century later. As Bette Davis is rumored to have said,

"In Hollywood, you either get cast in the same role over and over, or you don't work."

Inside Daisy Clover (Warner Bros., 1965)

"It's Mr. Raymond Swan. It's the movies!" Daisy exclaims. Some would argue that *Inside Daisy Clover* is one of the best movies about the film business.

Natalie Wood is Daisy. Wood is twenty-six, playing a fifteen-year-old. Her mother is played by Ruth Gordon and Mr. Raymond Swan is played sternly by Christopher Plummer. Roddy McDowell is Swan's studio yes-man and Robert Redford appears as the glamorous new golden-boy actor, set up to be Daisy's love interest until she discovers he is gay.

Oh yes, this film is "everything Hollywood." It tells the bittersweet story of a young girl who gets caught up in the adventures of becoming "America's Little Valentine," a persona created by Swan Studios. The manufacturing of a star is outlined here step by step. At the beginning of this star-saga, Daisy is living on a pier with her mother, whom she calls "Old Chap." The Clover family is dirt-poor, as Mr. Clover "took a powder" seven years ago. Daisy sings in a booth on the pier and is discovered by one of Mr. Swan's henchmen. Daisy, a rough and ready tomboy, is taken to the studio and measured up. Mr. Swan likes what he sees and takes Daisy in to be cared for by the studio, putting Old Chap into a nursing home. Daisy's greedy sister Gloria is made Daisy's guardian. Gloria eventually mismanages all of Daisy's money.

Daisy grows up in front of the camera, learning life's lessons as she evolves from street-smart kid to sexy starlet. As her star shoots heavenward, Daisy experiences her first drink, her first kiss, her first marriage and its dissolution, all in a dizzying blur. During the dubbing of a film that entails singing about the circus, she completely snaps, becoming hysterical while having to repeat the annoying circus theme over and over again. Finally, she suffers a nervous breakdown from the constant pressure of being made into something she isn't.

This 1965 movie reconstructs the hell-like restraints and constrictions of the 1936 studio system. It is a working-in-the-movie-business favorite.

The one-sheet (the poster used outside of movie theaters where the movie is playing) exclaims ". . . and a special world of thanks to all the slobs, creeps, and finks of the world. Without you my story could never be told . . ." Flamboyant and highly respected Hollywood writer Gavin Lambert, who wrote the novel this movie is based on and also the screenplay, set out to tell the tale of the road to stardom via his precocious heroine's own voice. Daisy's innocence is tested, as every actor's innocence will also be tested along his or her Hollywood journey to fame. Every actor has a Daisy Clover inside.

The Purple Rose of Cairo (Orion, 1985)

Woody Allen writes about what he knows—show business—and one of his best-loved movies is one that explores all the different levels of movie magic, *The Purple Rose of Cairo.*

During the Depression, Cecilia (Mia Farrow), a young working-class woman, goes to the movies to escape her life and to see her favorite matinee idol Gil Shepherd portray her favorite character Tom Baxter (both played by Jeff Daniels). One afternoon, the Baxter character, having had enough of the one-dimensional treatment he has been receiving at the hands of the screenwriters, walks right off the screen and into Cecilia's arms. When Gil Shepherd hears of his character's sudden materialization and independence, he and his agent worry about the effects this action could have on his career. It is clear that Gil has to stop his character before the character stops him . . . so Gil flies to New Jersey to talk to Cecilia.

This film is alive with the feeling of working in the industry. First, the fictional character is given dimension, and begins acting independently. The scene when Tom Baxter asks Cecilia what she really thinks of him creates a situation that doesn't happen often in the real Hollywood. She tells him the truth, and the truth is that he is not a great actor. Next, the actor Gil is forced to control his character, an interesting thought that most actors would probably love to act upon if they could write their own scripts. And third, here is a movie that becomes part of reality; it meshes with the real life that surrounds it. It echoes Buster Keaton's *Sherlock, Jr., Last Action*

Hero, and a little of *Sullivan's Travels,* as the character crosses over into a special world (actually, reality) and learns from that journey.

"In your world, things have a way of always working out right," Tom is told. And this is usually true; in the world of make-believe, things do seem to work out somehow. However, real life in Hollywood does not necessarily mirror the fantasies Hollywood is so good at creating. In fact, most of the time Hollywood is more like reality than real life.

Honorable Mentions of the Thirties

The Cowboy Star (Columbia, 1936)

Won Ton Ton, the Dog That Saved Hollywood (Paramount, 1976)

Under the Rainbow (Warner Bros., 1981)

Sunset (TriStar, 1988)

As the talkies took over during the late twenties and into the thirties, many silent actors didn't make the cut. One film, *Won Ton Ton, the Dog That Saved Hollywood* is the story of the rise and fall of a famous dog movie star, a satire on the life of dog star Rin Tin Tin. This movie features more than seventy silent film stars in cameos. As you watch these cameos, the fact that fame is difficult to hang onto really hits home.

The *Cowboy Star* explores the plight of the American cowboy from the time of the turn-of-the-century to the mid-thirties. Not only were there movies being made about these legendary characters, but the real persons themselves were often still alive and hanging out around the Hollywood area. (For another look at the movie-cowboy icon, see *Hearts of the West* in chapter 10. There is also Blake Edwards's *Sunset,* where Bruce Willis plays silent film star Tom Mix and James Garner is the aging marshal Wyatt Earp. This movie explores their work as a team on the back lots of Hollywood as they solve a murder mystery.) And finally, *Under the Rainbow* is the story of the special extras, the little people, a.k.a. the Munchkins, who were called to be part of a film titled *The Wizard of Oz.* Their unusual rise to fame is explored herein, and their theme is "There's no dream too big and no dreamer too small."

THE FORTIES

Two films of the forties offer realistic insight into the life of the aging actor and the trials and tribulations of the twilight of this life. *Dancing in the Dark* is featured here; *The Great Profile* is discussed in chapter 2.

Dancing in the Dark (Twentieth Century Fox, 1949)

Aging leading man Emery Slade (William Powell) could use a sense of humor. Arrogant and cantankerous (and apparently that way for most of his career), Slade is down on his luck during the post-World War II years, avoiding his landlady because he's behind on rent payments. One of his former colleagues from earlier, richer times runs into him at the Chinese Theatre and, seeing how depressed Slade is, makes a plea to the Motion Picture Relief Fund to help him. Slade gallantly refuses but is soon called upon by a studio exec that remembers his work—not to act, but to be a talent scout. The exec is banking on Slade's old association with a well-known actress, hoping that Slade will convince her to star in a hot property. Slade, however, discovers a new actress—someone far better for the part, and along the way learns of a new path for himself also. He's a sly old fox, but one that can be taught new tricks. The real-life lesson to be taken here? You never know when you'll be called upon to do an ancillary job in the motion picture business, and that job might take you where you wanted to be in the first place.

THE FIFTIES

The fifties are a time when all of the movies about Hollywood look pretty much like a gelatin-print black-and-white photograph. That glossy shine rings through most of these films, adding to the depths of the shadows formed by the sun. Los Angeles looks glamorous in black-and-white. Considering how much color was used in many of the feature films of that decade, perhaps the makers of the films about films shot in black-and-white to subdue the stark realities of the industry that were too jarring to be seen in their natural color state. Using black-and-white film was a technique that made the content unique and privileged. It was used as a way of celebrating the special world only the filmmaker himself really knew, the insider's view askew.

Sunset Boulevard (Paramount, 1950)

Perhaps the seminal movie when speaking of movies made about Hollywood, *Sunset Boulevard* is a perfect film. Here is the story of a silent film star and a silent film director who come together on screen to tell a classic story. This film is also discussed in chapters 4 and 10 of this book, but here, in the Actor section, the element of reinventing oneself is what is of interest. The life stories of the performers who appeared in this movie say almost as much about the trials and tribulations of aging film stars as does the film itself.

Erich von Stroheim (who plays Max von Mayerling in the film) was born in Vienna in 1885. Little is known of his early life until he immigrated to Hollywood working as an extra in *Birth of a Nation.* He then became an assistant to D. W. Griffith on *Intolerance,* and later a successful silent-film director in his own right. He emerged as an auteur, selecting and producing only the films he wanted to do throughout his career. In 1928, *The Swamp,* starring Gloria Swanson (Norma Desmond in Sunset Boulevard), began filming. There were difficulties on the set. Dollars were being spent unwisely, and soon von Stroheim disowned the film. It was finally released as *Queen Kelly* in 1929; however, it never achieved box-office success. Von Stroheim was then reduced to writing screenplays and acting in other people's films— fifty-two films, to be exact, between the years of 1934 and 1955.

Swanson, on the other hand, arrived in Hollywood in 1913 from Chicago, was spotted as a photogenic young beauty, and her career took off. She was thirty years old when sound arrived, and she embraced the talkies boldly. In 1950, Swanson was fifty-two, von Stroheim sixty-five, and with their history, the brilliant director Billy Wilder was the only one who could bring them together again. To complete this triangle, the young actor Bill Holden was hired to play the innocent out-of-work screenwriter Joe Gillis. Essentially, both the silent-film actress and the silent-film director reinvented themselves and accepted the changes in their industry. The result was the opportunity of a lifetime to play roles that would forever be part of cinema history.

In real life, Swanson and von Stroheim morphed into the new roles of Norma Desmond and Max von Mayerling. In addition, the on-screen characters also adapt and deal with the change the new screenwriter's presence offers them. For instance, when Desmond is watching one of her old movies, the footage is from *Queen Kelly*. Imagine von Stroheim's response to having his ruined masterpiece, a film he wanted to deny making, being utilized in this 1950 presentation. No one knew at the time that it would live on in the history of film. The Hollywood tale is one of tragedy, yes; however, before the movie ends, the discussion of adapting oneself to the changes in the industry permeates the piece.

The film presents an interesting contradiction. In real life, Swanson and von Stroheim adjusted well to their roles in this movie, even though those roles were not what they had previously done in Hollywood. Holden's Gillis, however, isn't able to adjust to being an aging star's gigolo, and Swanson's Desmond won't admit that actors need speaking roles in films. Both characters are forced to make choices about adapting their careers to the situation at hand. Ultimately, the movie is about the tragedy of being unable to move forward, as Gillis renounces his normal life and Desmond kills her gigolo screenwriter.

Desmond's pronouncements of "I am big, it's the pictures that got small" and "No dialogue, we didn't need dialogue, we had faces then," are famous. Norma Desmond held onto the past; in contrast, Swanson knew exactly what to do and how important it was to move on from the silent movies to the talkies. Many careers were over at that point, but not the ones that adapted.

Dreamboat (Twentieth Century Fox, 1952)

Another story of two silent stars and the career choices they made can be seen in *Dreamboat*. This 1952 drama features a Mr. Chips type, Thornton Sayre (Clinton Webb), and an aging glamour queen Gloria Marlowe (Ginger Rogers.) At the height of the silent-film era, they made movies together, and Sayre—formerly an actor and now a professor of Latin and English

Literature—was a Valentino-type named Bruce Blair. When her colleagues, who have seen her father as Bruce Blair on television, laugh at Sayre's young daughter, she runs to him to learn the truth. He confirms his past as a Lothario and insists that the only woman he ever loved was her now-deceased mother. Sayre finds himself in even more hot water as the school board is up in arms with his past as a movie star. It seems the recent broadcasts on television, the new medium that brought entertainment into everyone's living room, have brought much attention to their school. The only person delighted with the TV broadcasts is Gloria, who is happily raking in the money as she receives a second chance from her public.

Sayre is left no choice but to go to New York to challenge the head of the TV network, Sam Leavitt (played by Fred Clark, who shows up in *Sunset Boulevard* and *Hollywood Story*, playing essentially the same type of character), to stop airing the old movies. He is also reunited with Gloria, who plays on his emotions and is basically just a conniving witch. Nonetheless, Sayre is dead-set on "consigning Bruce Blair to oblivion"—which he manages to do when he wins a court case against the network.

Sayre returns home to learn that he has lost his job, but, ironically, he is saved by the very source he wanted to disregard. Hollywood offers him a new picture and a new contract. He accepts, and eventually regains his place on the silver screen. Gloria has the last word when they both find themselves back where they started.

Overall, this film provides an amusing look at the then-new medium of television and its use of product from the then-considered-ancient movie industry. Showing old, silent movies within the framework of the new, smaller medium that spoke to audiences right in their own living rooms was a whole new way of looking at the history of film, a bridging of the two generations, so to speak. Many careers were revitalized with the onset of the new medium of television.

All About Eve (Twentieth Century Fox, 1950)

Hailing from Wisconsin, little only child Eve Harrington (Anne Baxter) found that make-believe filled up her life more and more as she grew older, and the unreal seemed more real for her. "When you're a secretary in a brewery, it's pretty hard to make believe you're anything else. Everything is beer." The manipulative Eve slowly climbs the ladder to success as she becomes the secretary to famous actress Margo Channing (Bette Davis.)

Yes, this film is set in the world of New York theater, but save for the location, the story is a classic tale of ruthless ambition. For that reason alone, it is worth the watch to see this *seemingly* idealistic, dreamy-eyed young woman and the age-obsessed actress fight it out scene after scene.

Eve, the dedicated assistant with an agenda, is a top-notch performer. She worms her way into Margo's life through lies, deceit, and calculated schemes. She makes a play for Margo's husband, Margo's upcoming stage role, and Margo's entire life, as Margo, who has been too busy concentrating on her own insecurities and worrying about aging, slowly catches on. Eve's cunning illustrates that ambition is not always partnered with honesty. This film should be seen by those who strive to understand what happens when passion becomes an obsession.

The Star (Twentieth Century Fox, 1952)
The Goddess (Columbia, 1958)

Both of these films have a bold black-and-white style and both feature strong women in the leading roles. *The Star* features a "usedta-been" Margaret Elliot (Bette Davis). Margaret's career is over; she realizes she can't think beyond a script. (Perhaps in the world of make-believe movies, the character of Margaret Elliot is Margo Channing after Eve Harrington took away her shining star.) She's an Academy Award–winning actress, who is unable to get work past a certain age (and this most definitely still exists today). She's upset, she's out of sorts, she's gone berserk when she takes her Oscar for a drive down Sunset Boulevard—while drunk (the Oscar used in this scene is Ms. Davis's real Oscar). She knows she must sober up, and manages to get a job at the May Company, a department store on the corner of Fairfax and Wilshire. When she applies for the job and is asked if she has any experience, she snaps: "Four and a half years at Marshall Fields, lingerie department, what else is there to do in Chicago?" She leaves the job when two customers recognize her as a fallen star.

With bad investments and a daughter (Natalie Wood!), who is being taken care of by her ex-husband due to her own lack of finances, Margaret strives to get her life together. She meets a young producer whose heart is set on making a film about an actress who has put her career before happiness. She realizes she is this actress, and returns to her life, making a vow that things will get better.

The Goddess, starring Kim Stanley and Lloyd Bridges, is hedonistic and sexual. This film is one of the first to explore the fate of a young actress who becomes exploited in a pin-up, Playboy-bunny kind of way. Kim Stanley's character is beautiful as well as neurotic, and was rumored to have had been based on Marilyn Monroe.

Here again is the tale of small-town girl Emily Ann Faulkner who arrives in Hollywood and achieves her dream—all the glamour, success, and money she could ever think of. She is soon unable to handle the fame and experiences a nervous breakdown that leads to a life of drug and alcohol addiction.

The theme of this screenplay is reflective of fifties' sentiment. It starts the movement toward telling the truth behind the veil, to show a new element to fame—the dark side. Success as a goddess screen star can be hell. Prior to this, most of our other actors-to-be have pretty much returned home to the Midwest instead of turning to bad habits. In the fifties, things change dramatically for star wannabes. Movies begin to face the music and tell the truth.

A Star Is Born (Warner Bros., 1954)

The 1954 version of this often remade tale starring Judy Garland as lead ingenue Esther Blodgett and James Mason as the doomed Norman Maine is perhaps the best known version, better remembered than the previous 1937 version, the little-known *What Price Hollywood?*, and the 1976 Barbra Streisand musical version. In this film, Esther Blodgett is not a young woman hankering to be a movie star but a professional entertainer. She already has a little bit of fame singing in the Hollywood area, and it is Norman Maine who propels her to a bigger fan base. They meet at a gala star-studded benefit show held at the Shrine Auditorium. The love story ensues and Maine meets his tragic fate, just as he does in the other versions.

Perhaps it is Judy Garland's stellar performance, a performance that mirrors her own real-life experience with fame, that the mid-fifties audience embraced and other audiences continue to delight in to this day. Her enthusiasm coupled with insecurity, as the actress and her character deal with stardom, shines throughout the narrative. She is the quintessential wannabe-turned-star. James Mason's strength and calm help to give his singing female lead her stable foundation, and the psychological interaction between these two characters remains etched in the memories of most consummate American-movie lovers.

There is one scene that is necessary—very necessary—viewing for all individuals wishing to work in show business. That scene is when Maine asks Esther what her dream is, and she replies that she would like to have a number-one record that would be played on every jukebox across America. He tells her: "You're better than that, you're better than you know. . . . Don't settle for a little dream, go on for the big one." Her dream isn't big enough.

THE SIXTIES AND SEVENTIES

In the fifties, the harsh reality of fame reared its ugly head. Actors and actresses turned to alcohol and drugs, admitted failure, and the decade was awash in really bad scenarios of this business called show business. The sixties would continue to add to this reality-based theme . . . and even go a bit farther than anyone making films about working in the industry would have imagined up until this point. And finally, in the seventies, tacky tabloid elements take over completely and give birth to movies that exploit the movies-about-moviemaking genre.

What Ever Happened to Baby Jane? (Warner Bros., 1962)

It's a good-sister-versus-bad-sister romp through hell as these two old biddies worry about losing their home. These two have had a twisted relationship since childhood, but now they find themselves doing the best they can to manage their Hollywood Hills lifestyle. Neither one has worked in years, and Blanche (Joan Crawford), a former actress, is in a wheelchair, the result of a car accident many years before. Jane (Bette Davis), a former child vaudeville star, becomes Blanche's reluctant caretaker.

Jane decides to stage a comeback to make money. The comeback doesn't work, so Jane decides to drive her sister out of her mind in a series of grisly incidents (like serving her her pet parakeet and roasted rats for dinner.) When Blanche confesses the truth about the accident that had left her para- lyzed, Jane decides to take Blanche to the beach and bury her alive. Jane becomes even more delusional as the police arrive and a crowd gathers. The final scene is reminiscent of *Sunset Boulevard,* as Jane welcomes the crowds and begins to sing and dance for her public once again in her life, this being, of course, her final performance.

This film was released in 1962 and was a huge box-office success. At the time, many considered it to be a horror film. Now it plays like a depressing piece from one of Hollywood's dark pockets of history, or high camp. The movies that show us the reality of silent-film stars wasting away in their mansions (see also *Sunset Boulevard* and *Inserts*), do not present a pretty sight. Some silent-film stars never worked again (and are sometimes referred to as "the wax works," a phrase originally attributed to director Billy Wilder and illustrated by Norma Desmond's card-playing friends in a scene in *Sunset Boulevard*), and others made the technology work for them. Just think of all the other faded stars populating the Hollywood Hills.

The Patsy (Paramount, 1964)

Entertainer Jerry Lewis creates a perfect scenario to explore the absurdities (and truths) of working in show business. A famous film star is killed in a plane crash, leaving his production team—including his publicist, producer, director, writer, and their secretary—out of work. The team is accustomed to a certain way of life, so they decide to find a "patsy," a new talent, to present to the world. At this point, the hotel bellboy Stanley Belt appears and is per- fect patsy material. The following ninety minutes are spent grooming the nerdy and awkward Belt into a famous and wildly successful movie star.

This movie looks like *The Errand Boy* meets *The Nutty Professor,* as Morty S. Tashman becomes Buddy Love.

The movie was of course meant as the usual Lewis physical comedy vehicle, but in retrospect, it can now be seen as a study of how the movie business operated in the sixties. The production team displays a sixties version of what the star-making machinery was like as the movie business came into the modern age. The birth of early publicity stunts and the elements of product placement and merchandising are discussed and implemented within the movie's comedic scenarios. Lewis was one of the first stars to have an exclusive deal with a studio—Paramount. He was guaranteed an outlet for his movies via Paramount Studios. In other words, he was so popular that just about any movie he made would be a success at the box office—and Paramount knew that, so they signed him on exclusively.

Watch this movie if only for the segment about publicity. Here, homage is paid to all the beauty parlors and barbershops across the country, for they are the "center of American cultural exchange," and it is there that the buzz is planted about Stanley Belt, the new kid extraordinaire. This montage shows how simple it was to plant information in the right person's ear so the buzz would be generated among the right demographic. "Creating the buzz" is a phrase used today to describe the generating of interest for a project that may or may not be worthy of the attention—just the mere fact that it is getting talked about is what publicity is all about. Through the escapades of *The Patsy* and later *The Errand Boy,* Lewis demonstrates his knowledge of the show-biz experience, and both movies are tributes to the industry.

Hollywood Boulevard (New World, 1976)

This film is pure exploitation. This was the seventies, and independent filmmakers were making films left and right with no rules, no control from a studio system. It makes sense that low-budget king and queen Paul Bartel and Mary Woronov would do a spoof on Hollywood and its wicked ways. Opening with newbie Candy (Candice Rialson) arriving in Hollywood straight off the bus, a montage of the footprints at the Chinese Theatre,

Schwab's Pharmacy, and a shot of her sitting on the Hollywood sign (something that isn't done very often, if ever) fills the screen. Candy even reads a paperback titled *How to Break into the Movies.* Kitschy music adds to the surreal scenes, as clichés and stereotypes of agents, directors, stars, and most of the crew act out in over-exaggerated scenarios. The movie is primarily an excuse to show lots of tits and ass, nudity, silliness, and stupidity. When a couple of street toughs ask her to take part in a bank robbery, Candy actually believes this will help her become a star. She's asked to drive the getaway car and keeps looking for the cameras. There aren't any, of course, and they get away—she is lucky to be alive. This is clearly Bartel and Woronov's personal homage to the industry they have both exploited—and have been exploited by—many times before.

Honorable Mentions of the Sixties and Seventies

The Comic (Columbia, 1969)

The Oscar (Embassy, 1966)

Valley of the Dolls (Twentieth Century Fox, 1967)

Play It as It Lays (Universal, 1972)

In *The Comic,* veteran show-business insider Carl Reiner utilizes his television lead actor Dick Van Dyke in a show-biz yarn about Billy Bright, a silent-film star who tells his life story on the occasion of his funeral. The only reason to watch this film is to understand that stars have normal lives too . . . and Bright's life without the camera is ordinary beyond belief. *Valley of the Dolls* and *The Oscar* are offerings in the vein of the camp variety. In *Dolls,* three young women experience Hollywood in the sixties via abusive love affairs, potential jobs in porno movies, popping too many pills, and numerous scenes of extraordinary female hysteria. *The Oscar* focuses on one egotistical no-talent actor who is forced to work in television after his film career dies. And finally, *Play It as It Lays,* a movie based on Joan Didion's novel of the same name, is a product of seventies American cinema gone amok. There is a small storyline regarding a gay producer and a troubled actress that seemingly has no beginning or end. The real star of this movie is the Los Angeles freeway system, which is featured heavily during and in between many scenes.

THE EIGHTIES AND NINETIES

While everyone was busy making money and being self-centered in the "me" decade, Hollywood took a break from itself. There are not that many films in the eighties concerning the topic of stardom. Perhaps the last two decades' worth of bad films on this subject made the studios weary of it. Rightfully so. Fast forward to the late eighties and all of the nineties, though, and you'll find a good sampling of films that dive right into the usual star-making madness.

Hollywood Shuffle (Samuel Goldwyn, 1987)

Hollywood Shuffle is a movie about becoming a star, a feisty tale of a young man's struggle to act in and direct his own movie, made by the actor-director Robert Townsend. *Hollywood Shuffle* took two and a half years to make. It matured from a series of vignettes into a full-length film. There's a parody of Siskel and Ebert in "Sneakin' into the Movies."

A funny bit at the Winky Dinky Hot Dog Stand, and the lead character Bobby Taylor (Townsend) auditioning for Jive Time Jimmie's Revenge where he is told he is not black enough. Townsend himself has said that when that happened to him in real life, he made the decision to make his own movie. The result is this realistic look at a black actor's life in the business.

In many ways, this film is important, not only because it was independently produced but also because it was ahead of its time. Since the mid-nineties and through the turn of the twenty-first century, there have been cries for more actors of color and more roles for them. *Hollywood Shuffle*, albeit a comedy, had brought initial attention to this very important topic in the industry.

This was also one of the first independent films to have been financed by credit cards. With an alleged budget of $100,000, Townsend made it known as part of the publicity for this film that plastic provided the funding for his project. The result was a successful film that continues to resonate with all audiences.

Postcards from the Edge (Columbia, 1990)

Carrie Fisher's best-selling novel is a perfect movie about being a star—or, in this case, an up-and-coming actress who is a star's daughter. It is also a realistic look at the less glamorous aspects of being a star, the realities one faces growing up in the shadow of the Hollywood sign. As the nineties are born, not only has the "Esther Blodgett" prototype character blossomed, she now has a number of additional issues to deal with, not the least of which includes dealing with her aging show-biz mother. Meryl Streep stars as Suzanne, an actress who is dealing with the usual ups and downs of working in the industry. She has had a history of drug dependency. She is sarcastic yet vulnerable, and she is caught in the web of Hollywood.

Throughout the film, relationships are examined as closely as, if not more closely than, Hollywood itself. A harsh mirror is held up to Suzanne's work in a rehab center. She must face her domineering mother and suffer numerous humiliations at the very hands that raised her. When she shows up for work on a B-movie set, she is forced to give a urine sample. A producer takes emotional advantage of her, and her agent betrays her by running off with all of her money.

Ultimately, Suzanne overcomes all of these pitfalls, learning that there are simple moments in life beyond the glamour and the praise. She finally achieves a perfect acting performance and realizes that that moment is more intense than the drugs, sex, glamour, and acclaim that she sought earlier. The simple joy of friendships with her mother and others bring her much more now that she is soberer and wiser. This was the elixir Bette Davis was looking for in *The Star*.

Last Action Hero (Columbia, 1993)

One of the biggest box-office bombs in the nineties was Arnold Schwarzenegger's film *Last Action Hero*. This $80-million adventure is reminiscent of *The Purple Rose of Cairo* in that its main character steps off the screen and into reality to complete his journey. This time, the character Jack Slater (Arnold) stars in *Jack Slater 4* and has a sidekick, young fan Danny Madigan (Austin O'Brien). This film is a meta-movie. That is, a movie within a movie; and that movie within makes fun of itself and the action genre. The result is a film that goes beyond the original genre. (The *Scream* series is another example of this type of film.)

The story warns of the pitfalls of the genre but celebrates it at the same time. This movie shares an important moment with *The Purple Rose of Cairo* and Buster Keaton's *Sherlock, Jr.*, when the character walks off the stage and into reality, or from reality into the movie. These are beautiful scenes of going in and out of the media, again walking that fine line between fantasy and reality. This is, of course, what films are supposed to do. Mostly, the story is a commentary on the big blockbuster genre. Quite a few of these formula blockbuster films use the same recycled plot, with a vague, generic hero at the center. The hero performs death-defying feats that would kill any other man, while at the same time saying something witty. The world of the formula blockbuster is an idealized one: the sun always shines, everyone is gorgeous, and good always wins out in the end—after tons of mass destruction.

Last Action Hero never condemns or condones the action genre; it just heightens everyone's awareness of it. Of course, the genre has run itself out, and by the mid-nineties, surely by the turn of the century, the big-name action stars' plots heavy with explosions, special effects, and rehashed story-lines became history. In the aftermath of the September 11 terrorist attacks, the action genre pales in comparison to real life. Action movies will need to reinvent themselves to reflect their times.

Swingers (Miramax, 1996)

The movie that influenced the first generation of media moguls of the twenty-first century is *Swingers*. Writer-actor Jon Favreau relocated from New York to L.A. to pursue a career in acting. This was his first script. This story was his story, his experiences at Hollywood casting calls and in the nightclub scene. Before long, he had pages of scenes that described a group of people hanging out in cool places, talking an absurd language filled with linguistic slang reminiscent of the early Hollywood martini-lounge scene— "You're so money, man!" "I got digits, baby." Through this young, desperate, and struggling actor's point of view, the world of twentysomething retro-Swing dance kids and the newly revised Cocktail Nation of cigars, scotch, saucy polyester garb, and golf clubs comes to life.

Favreau's character, Mike, has just gotten out of a six-year relationship and he misses his ex desperately. Buddy Trent (Vince Vaughn), is a ladies' man, wanting Mike to meet some new "beautiful babies." Mike, Trent, and their friends roam around the dark clubs of L.A. by night, and by day go to auditions. These hipster wannabes accurately reflect what it means (and takes) to be successful in show business at the century's end.

As a commentary on the acting profession, this film looks at the changes in the industry and the fact that actual acting ability is not really needed in many of the filmed scenarios of the nineties. You need something more, something that sets you apart and gets people talking about you. If you don't have "It," you might as well give up. And yes, the "It" factor has been around since Hollywood began, but as Mike and his buddies discover, sometimes all someone needs is "It" and nothing else. Entire careers are made on "It." From the thirties through the sixties, actors made attempts at learning how to act by attending acting classes—that's not necessary anymore as long as you have "It."

With swarms of actors everywhere in Hollywood (remember the one-in-one-hundred-thousand chance way back in the twenties and *What Price Hollywood?*—it hasn't changed), it's no surprise that no one is surprised when you give the answer "I'm an actor" to the question "What do you do?" Hollywood has become this pseudoenvironment where you must put forth your best effort to get any attention at all, and this movie confirms it.

Notting Hill (Polygram, 1999)

Speaking of ordinary people, the last film of the twentieth century that deals with being a star involves just that—William Thacker (Hugh Grant), an ordinary bloke who runs a travel-book store in the Notting Hill section of London. Enter famous film star Anna Scott (Julia Roberts), and an attraction is instantaneous. William accidentally spills orange juice on Anna and invites her back to his place to clean up. She kisses him in thanks, and he says "It was nice to meet you—surreal, but nice" and later rents her movies, watching them with his wacky roommate. William can't believe his good fortune—he has kissed a movie goddess.

Reality overlaps again when Anna calls William. He finds himself in the middle of a press junket at a local hotel and gets in to see her, heavily guarded as she is. "It's the sort of thing that happens in dreams," he tells her, "not in real life. It's a dream to see you again. What happens next in the dream?" And so, this enchanting love story echoes what it is like when normal mortals fall in love—for the experience of love often feels like a dream. But in this case, the mortal is truly falling for the immortal. *Notting Hill's* surreal scenario is one that echoes the *Star Is Born* prototype in nineties fashion. As the two meet again and again, prompted by Anna's desire, they consummate their love. Anna tells William about actress Rita Hayworth, who said, "They go to bed with Gilda and they wake up with me." (*Gilda* was one of her most famous film roles.) Anna muses, "They go to bed with the dream and wake up with reality." And so, even actors and actresses are real people, and in this case, the actress-goddess becomes more real due to her relationship with the mortal. *Notting Hill* is a sweet story of stardom, one that nicely blends these two worlds and sets the stage for the new century.

ACTOR WRAP

Throughout nine decades of movies about being an actor, we've seen the slow changes in the way the image and the lifestyle of stardom is portrayed. The early extra girls and "Mertons" guided us through the Chaplin era and the silent-film epoque. Next we saw the emergence of the confident star climbing her ladder to fame through the *Star Is Born* myth. By the middle of the century, the truth behind the curtain, so to speak, the behind-the-scenes reality emerges.

Being a star really hasn't changed through the years. You need as much passion for your dream as all those who flocked to Hollywood a hundred years ago. Just make sure the dream is big enough to carry you through the entire century.

IN REAL LIFE

For an insider's look at what it's like currently like to be on the actor beat, here are two reports from the trenches.

Young Thespian Sammy Shorewood Speaks Out

Sammy Shorewood has been working in Hollywood for the past three years. He's been to almost every audition for the "young, quirky, comedy type." Tall and blond, looking like a cross between Dennis the Menace as an adult and that lead actor from *Dawson's Creek,* Sammy has some advice for young actors. His visits to Hollywood casting couches and cattle calls are recent, and here's what he has to say.

> The moment you sell out is the moment your love for creating is replaced by a love for being admired for creating. Focus on the process, not the result. If you focus on the work and love the process, then the three-picture deal will fall into your lap. Conversely, if you crave the result (the car, the fame, and the power) more than the process, you will eventually realize you are a fraud, lose the will to create and become one of the many functional alcoholics working in the industry today. You don't want to be them. They are bitter, jaded people with empty eyes. It takes years. Don't get jaded. It's all in the eyes.

When asked if success is just a dream or something that is really tangible and real, he responds:

> If you are a confident, motivated, and professional creator working in the industry, the odds of success are in your

favor, and anyone who tells you otherwise is either a) regret-
ting a dream they themselves didn't follow (REGRET), or
b) trying to protect you from the greatest adventure of your
life because they want your story to be as safe and pre-
dictable as their own (JEALOUSY).

Beau Diamond Grabner: Extra Work Extraordinaire

Beau has experienced almost every entry-level job in Hollywood since his
arrival here from the Midwest straight out of college. Energetic, enthusiastic,
and ever-happy, Beau knows his movies . . . and he's wanted to be part of
them ever since he can remember.

How did you get work as an extra?

I set myself up as an extra by simply going to one of the many extra agencies in town and signing up with them. There is an office, and they call you
if you fit what the production is looking for, and check your availability. I
would work from zero to seven days a week, depending upon what was
shooting. I was typecast as a "white guy"; it's all about typecasting, and
being a white guy, I filled a lot of types and mainly did crowd work or guy
sitting at a table.

What's life like on the set?

You pretty much do what you are told and sit around and wait for when
they need you. I recommend bringing a book. The pay varies from one
production company to another, usually from $40 to about $100 for eight
hours, and then you get overtime on some shoots. The pay is usually not that
good, but you get to meet other actors and interesting people. The agencies will
give the more experienced extras the higher paying jobs, so yes, experience matters in some cases, and in other cases it is just luck, how much you get paid. I
also recommend having another job that is flexible and that you can do at home
or at night, like transcription.

Would you do this on a long-term basis?

Both socially and professionally, working as an extra can lead to either making a good friend or finding a job opportunity. I would not do it full-time. Others love it, but it's not for me. Aspiring actors should do it for a while, to meet people and get some experience in front of the camera. Production people should do it once or twice, just as an experience. It helped me to understand how to work with both actors and extras, once I started working in production.

CHAPTER 2

AGENT-MANAGER

CREATIVE CAREERS IN HOLLYWOOD

STATUS

DURABILITY: Keeper, as long as you don't burn out.

LENGTH OF STAY: A good five to eight years, at least, and perhaps an entire life's career.

FOOD-CHAIN VALUE: High.

UPWARD MOBILITY: Why would you? You've already got power.

DESIRABILITY FACTOR: Very high.

VACATION: Possibly, but never give up access to your computer, cell phone, and assistant.

SALARY: Venti Iced Mocha Coconut Macchiato.

HOW EASY IT IS TO GET THIS JOB: On a scale of 1 to 10 (1 being the easiest), 5.

PREREQUISITES: Being devoid of any sort of etiquette, knowing the proper use of profanity, being a graduate of How to Schmooze 101, and just generally knowing how to be an asshole.

There was nobody calling me up for favors . . . no one's future to decide.
—Joni Mitchell, "Freeman in Paris"

Agents. Most of them have a bad reputation for being assertive, pushy, bitchy, and determined beyond all means. This is a good thing. If you are a natural at sales and promotion, then being an agent or manager is the job for you. In addition, if you find yourself utilizing the art of schmoozing within your daily life, you'll find that this position is perfect for you. Perhaps you always knew how to manipulate your family and friends to let you get your way, and the word "no" is unacceptable to you on all accounts, well, then, welcome to the world of agenting.

There are numerous jokes about agents not being human, lacking any compassion, matching the energy of sharks, and never knowing how to use the phone to call their clients and keep them abreast of the work they are doing for them. Jokes or no jokes, the agents who use attack techniques, forcefully make their point, and consistently focus their energies on clients who can make the most money for them are the most successful. As an agent, you can enjoy a lucrative career with one or two clients, or a stable of different talents. As an agent, you can have power—the power to marry talent to a great script or to discover the next supercelebrity sensation, or negotiate the deal that will jump-start a pop-culture phenomenon. Agents and managers, while having a bad rap, can actually be quite creative, especially when they are able to see the glimmer of promise and the wealth of potential in the talent they sign. This chapter focuses on five movies that feature agents in primary roles.

SCHMOOZING AT ALL LEVELS

There are various kinds of agents with specialized levels. In addition, there are managers, whose tasks are slightly different from those of agents. Here's the breakdown within the Hollywood Food Chain:

Agents

Agents are the puppet masters of the business. They negotiate and they weave the fabric that keeps the Hollywood Food Chain together and flowing. They are a necessary evil, and most know to cooperate with their clients, producers, production companies, and studios. However, some do tend to play in the "outrageous child" mode and make impossible demands to make their point or get their way—either way, agents are important and ultimately very useful in getting things done in Hollywood.

Agents specialize in different aspects of the business, much like lawyers. And just as one would not typically hire a bankruptcy lawyer to handle a divorce, agents have their own areas of expertise. Expertise such as actors, directors, writers, producers, cinematographers, editors, and some crew positions. Representation means that the agent will handle all business transactions for the client, that is, negotiations regarding terms of contracts per project, and generally act in the best interests of the client in all business aspects. The specific types of agents are:

- **Talent agents**—represent actors.
- **Literary agents**—represent writers; both fiction writers and screenwriters.
- **Voice-over agents**—represent individuals who work in voice-over only.
- **Director or producer agents**—represent these principals and other crew positions.
- **Packaging agents**—agents who work for a large agency, who take a project and package it. They link clients from their agency, and their agency only, to one project; i.e., the writer, director, lead actors, and producer (or any combination of the above) are all represented by the same agency. This way the packaging agent keeps all of his agency's clients employed and the profits and percentages coming back into the agency.

Of the above-listed agents, each line can then be broken down into agents who represent specified talent for film and specified talent for television. As outlined on the Hollywood Food Chain, the industries of film and television, while similar, are actually very different in many ways. Specialized agents are needed for each industry.

Managers

While agents represent, negotiate, and handle clients, managers are like coaches. They will manage their clients' careers, giving them advice about long- and short-range career moves and strategies. Furthermore, the manager will then help the client implement these moves. A manager is like the orchestra leader of his client's career, organizing the career through various agents, publicists, lawyers, and perhaps a business entity.

THE TENPERCENTERIES

Agents generally work within agencies, although it is not mandatory to do so. However, as illustrated by the above list, if a project can be packaged within one agency, it is to the agency's advantage. There are large agencies that feature all the various departments of the industry listed above, and there are smaller agencies (generally referred to as boutique agencies), where agents specialize in just writers, just directors, just voice-over talent, and so on. Managers generally work on their own, or, again, within a smaller organization.

Agents take 10 percent of their clients' salary, while managers typically take 15 percent. Often, agencies are referred to as "tenpercenteries" for this reason. This is also the reason agents and managers strive to get top dollar for the talent they represent: the higher the client's salary, the higher their percentage of the pie. They are the hustlers extraordinaires of Hollywood. They live to keep the supply of talent in demand.

Agents and managers make the world go 'round—at least within the entertainment industry. The following films commemorate five individuals who have chosen to guide other people's lives—and get their fair percentage for doing so.

The Great Profile (Twentieth Century Fox, 1940)

The Great Profile is an obscure film made at the end of the great actor John Barrymore's career. Barrymore was known to be fond of drinking. In this farce, he lampoons himself portraying the famous actor Evans Garrick, who is also given to drink. Garrick nearly destroys the show he is hired to act in. This film is considered biographical of Barrymore. When Garrick's wife leaves and his agent quits, one Boris Mefoofsky enters the scene.

Mefoofsky, a.k.a. "Mafoo," is played by the eccentric character actor Gregory Ratoff, last seen in *What Price Hollywood?*, where he played a Hollywood producer with much the same energy as he uses in this film. Garrick's performance parodies himself, his profession, and his audience. Throughout the movie, the sharp, gusto-filled Mafoo matches Garrick's slick one-liners and challenges the grand thespian.

Mafoo is able to call Garrick on all of his crap and make the situation work. He cleans up Garrick's mistakes, often because in the end it will benefit him to do so. Mafoo's performance shows the loyalty of an agent to his client, which had not been explored previously in film. This early display of an agent's loyalty is a precursor to the agent-covered scandals in which many contemporary stars get entangled. There is little difference between Mafoo's controlling attempt to harness the drunken situation Garrick finds himself in and the modern-day spin management the agents of Shannon Doherty or Ben Affleck have to apply to tabloid reports of their sometimes-public scenes of drunken behavior. Mafoo is fierce, as are contemporary agents and managers, for they must protect their clients—their commodities and their bread and butter on all fronts.

Actors and Sin (United Artists, 1952)

There are two parts to this movie and it is the second half, titled *Woman of Sin*, that provides a look at one Orlando Higgins (Eddie Albert). Orlando is the quintessential agent. He's full of one-liners as he barges into his office, embraces and tongue-kisses his secretary, and sits down behind his large desk, phones a-ringing. She tells him an author is on the line; he responds, "I hate authors. I get sick any time I talk to one." Then he is told that a secretary has made a mistake and mailed a project to Empire Studios—she was supposed to send it back to the writer. Now the studio wants to buy the project, and he replies, "Go back and see if you can make any more mistakes. It's the only way to succeed in Hollywood." Orlando is soon introduced via the phone to one Daisy Marcher (Jenny Hecht), the author of *Woman of Sin,* the hot script. She tells him that she's in Palm Springs and can't be bothered. He dismisses it for now and goes to play tennis and swim.

For the fifties, Orlando has a nice life. In between his workouts, he manages to get a $75,000 deal for Daisy. The studio wants to meet the author. Orlando has to act fast, as they are now willing to make a $100,000 deal, which even by today's standards would be nothing to sneeze at. Daisy is finally ready to meet Orlando. When a nine-year-old blond-haired girl shows up, Orlando asks to see her mother. Daisy insists she is there alone and that she is the author of the hot script. She convinces Orlando of the fact, and soon he has the task of controlling an obnoxious yet delightful little girl.

Orlando brokers Daisy's deal, acquiring patience along the way. When Daisy writes her next piece, *Sea of Blood,* and the studio comments on its violence, Orlando screams at Daisy: "Why didn't you stick to sex?"

Actors and Sin is a gem of a film, perhaps ahead of its time. Nicely executed and sarcastically written, it is an accurate portrayal of the vocation of an agent. This is not a profession for the weak. *Actors and Sin*'s writer Ben Hecht describes an agent as "a ten-percenter, peddler of genius and beauty, full brother to the Headless Horseman—evasive, double-talking, irresponsible as a grasshopper, liaison officer between the Mad Hatter and the Three Little Pigs." Orlando is the epitome of Hecht's description, as Eddie Albert's laissez-faire, ego-filled portrayal of the lead character succeeds wildly. When released, this film created a huge brouhaha, because it allegedly ridiculed the film industry excessively. Hecht's version of his experiences within the industry struck a chord of guilt. These scenes were truthful—embracing all the lying, deceiving, and conniving found in an agent's day's work. It was a little too true for many in the industry at the time. All these years later, the portrayal of Orlando is quite on target. This is not ridicule so much as the truth.

Star 80 (Warner Bros., 1983)

Directed by Bob Fosse, *Star 80* is one of the darkest films analyzed in this volume. Produced in 1983, this is the true-life tale of *Playboy* model Dorothy Stratten's brief and tragic career. Slick small-town hustler Paul Snider (Eric Roberts) discovered Stratten (Mariel Hemingway) while she was working in a Canadian Dairy Queen restaurant. Roberts is pure entertainment as he revels in this sick puppy's weirdness, beefing up the way he walks, talks, and combs his hair—right down to the pencil-thin mustache that only adds to this guy's creep factor. Snider is an example of what not to do as an agent-manager. While it is good for the agent-manager to become involved in the client's life, in this case, Snider went too far—way too far.

As the beautiful Dorothy becomes a star well beyond the pages of the bunny magazine, Snider wants control of her body, her mind, and her money. He talks her into marrying him. At the same time, Dorothy begins to make the rounds in Hollywood, mingling with rich and powerful people. She meets real movers and shakers. They shun Snider and his small-town arrogance. In reality, he is nothing but a loutish and vile fanatic who becomes selfish and overpowering in order to keep Dorothy under his control. The results are horrifying, as his jealousy overtakes him and he kills her and then himself in a blood-soaked massacre. Ultimately, his ambition was thwarted by his envy, and her star shot down by his insanity. This is a gruesome film that illustrates the extremely dark side of managing a client's career.

Broadway Danny Rose (Orion, 1984)

In this movie, Woody Allen is the pathetic but delightful third-rate talent agent Danny Rose. Now, okay, this film does not take place in Hollywood, but it does tell the story of a dedicated agent, and it is the film that offers the most insight into the job of being an agent, so in this case, the borders of Hollywood extend to New York City.

A group of veteran show-business entertainers meet at the Carnegie Deli and tell their own personal stories about their involvement with Rose. The result is a charming look at the dedication and special relationships often found between agent and client, and for that reason, this film should be seen by anyone considering a career as an agent or manager.

Rose is not very successful as an agent—and his clients are not very successful either. Nonetheless, he feels the obligation to go out of his way to help his clients in their time of need. In this case, singer Lou Canova (Nick Apollo Forte) wants to make a comeback, but his mistress (who poses as Rose's girlfriend whenever Canova performs so Canova's wife, who is in the audience, will never know,) is being followed . . . by mobsters. The comedy takes off as meek, enthusiastic Rose deals with the adventure at hand, all the while pitching his clients with lines such as "My hand to God, she's gonna be at Carnegie Hall. But you—I'll let you have her now at the old price, okay? Which is . . . which is anything you wanna give me. Anything at all."

Danny Rose means well. His heart is in his job, a rarity for an agent. This is a sweet portrayal of the occupation of agent, a labor of love by one of the industry's grandmasters of comedy.

The Big Picture (Columbia, 1989)

This final selection is required viewing for anyone who wants to work in the industry. The film is discussed further in chapter 10, but for the purposes of this chapter, Martin Short's interpretation of a Hollywood agent is a near-perfect parody. Short's Neil Sussman is a fast-talking stereotype of the quintessential egotistical agent. Everything he says is babble, really, but it's delivered with slimy intimacy, in perfect, Beverly Hills, show-biz intonation. Nick Chapman (Kevin Bacon) won the highest trophy for his student film produced while he was attending the National Film Institute. Hollywood is a-buzz with Nick's name, he is the flavor of the day, and everyone wants to sign him. Enter one Neil Sussman, agent extraordinaire.

Martin Short, who is uncredited in this film, is first seen waiting for Nick to join him to have lunch at a fancy-schmancy restaurant. Short plays Neil as an ultra-effeminate middle-aged man trying to pick up another guy at the restaurant, even though he announces that "his wife and he" dine there often. Nick arrives and greets Neil with a respectful "Mr. Sussman?" to which Neil responds, "Mr. Sussman is my father and he lives in Miami

Beach, call me Neil." Neil goes on to admire Nick's eye color and opens the lunch with "Look, Nick. I'm not going to bullshit you. I don't know you. I don't know your work. I think that you are a very talented young man—and I'm never wrong about these things . . ." Neil is ready to sign Nick without even seeing his film, just due to the industry buzz Nick has received.

This is not merely a screen fantasy. If a student or indie-made film receives praise, that praise will spread like wildfire among the agent-and-manager community. Often, a young talent will receive hundreds of offers based on nothing but the fact that the rest of Hollywood is chatting about the project, so everyone wants to get their hands on it.

Nick is hardly able to get a word in edgewise, nor is he able to order any lunch as Neil continues his monologue. "If you decide to sign with me, you get more than an agent. You get three people (Neil holds up four fingers): an agent, a mother, a father, a shoulder to cry on, and someone who knows this business inside and out . . . and if anyone ever tries to cross you? I'll grab them by the balls and squeeze till they're dead." Neil, in all of his eccentricity, is actually offering Nick a pretty good opportunity.

Nick signs with Neil, and Neil does prove to be loyal throughout Nick's young roller-coaster career. Finally, when Nick wants to hold out for final creative control in a movie he's about to make, Neil warns Nick about being too demanding but supports him in his choice. Nick wins the opportunity to do the project—his way. Neil, who allowed Nick his freedom, yet remained by his side, is the winner. Overall, while Short plays this role with a campy attitude—"I've read almost all of those scripts almost all the way through"—he does in fact prove to be a good agent.

THE BEAUTIFUL YOUNG TURKS

Orlando, Mafoo, Danny, Paul, and Neil are agents who have successfully done their job. These celluloid agents (with the exception of Paul Snider) do their best to seek out work, cover up any undesirable information, and represent their clients.

There have been some extremely notable agents and managers throughout the history of Hollywood, such as Swifty Lazar, Abe Lastfogel,

Lew Wasserman, Charlie Feldman, and Mike Ovitz, to name just a few. Many have started at the lower rungs of the Hollywood Food Chain and worked their way up. Starting in the mail room is an ordinary route for many future agents. As the mail is passed around, they learn who is who and what is what. The basic knowledge of the agency itself will help the individual grow quickly as he moves upward. Many will find a mentor and follow in that agent's footsteps, often taking over some of the mentor's clients. With a lot of ingratiating and manipulative behavior, the rookie will continue to make contacts and engage in overall self-promotion. Then the Young Turk will take over and become an agent in his or her own right. It is desirable to become the meanest, most feared agent around, intimidating your opponents (all other agents vying for the same hot properties), thinking as a warrior, and knowing that being popular with your clients is not one of your goals.

Being an agent is the true realization of living in the fast lane. Many burn out, either becoming workaholics or resorting to drugs. This is not an easy job. This is not a job that makes a lot of friends. It requires a dog-eat-dog type of energy, and if any vulnerability, weakness, or lack of self-confidence rears its ugly head, you can bet the individual will fail and find it difficult, though not impossible, to climb back up again to his lofty heights. Nonetheless, agents and managers ultimately are the kings behind the business. They make deals that in turn make money. It's a win-win situation and it works well.

IN REAL LIFE

Young manager Chris and junior agent Heather offer up some thoughts on drive, determination, and solid confidence.

"I'll Manage Your Career" Chris

For the past few months, Chris, in his mid-twenties and hungry for success (he was once quoted as saying "I don't think $250,000 is a very high salary, sounds about right as a bonus, but not the base salary"), has been operating his small management company from his desk as an assistant at a major network. (For reasons you can certainly understand, he would like to remain anonymous.)

"As a kid growing up in L.A., I had an uncle in the business. He had more exciting things going on than my dad or anyone else I knew. His weeks were full of premieres, dinners, and speeches. His winters were spent on the islands and summers in the mountains." Who wouldn't want that type of life?

And so, Chris modeled himself after the uncle, gaining work as an assistant at a busy cable-network desk. Soon, he realized that he could spot talent as well as the pros that worked on the same floor as he did. "I think I have an eye for talent. I don't think I can make any prediction about what audiences like or don't like, but I do have an eye for talent." From that point, he signed his first client, and then his second, and third. In-between scheduling meetings and airline trips for his boss, he sells his client's projects.

"Competition plays a huge part in every aspect of this business. If you don't have competition, you have no reason to excel. I think everyone on the creative side of this business has a responsibility to raise the bar with every project. If we keep on making progress creatively, we will continue to grow intellectually."

Obviously, Chris is working in an environment that allows him to nurture his clients, as his personal management company is very small.

> Yes, I have been fortunate to take some exciting young clients out of the clinches of the larger management companies. Make no mistake, when the writers get hotter, the vultures will wine and dine my impressionable clients. I think the advantages of being small are as follows. . . . I don't answer to anyone other than my clients. I don't have a boss forcing me to service clients I think are hacks. I may join a large agency but I would rather just own one.

Heather Wilder on Doing the Agent Job Right

Heather is twenty-five and has just spent the last three years at one of the top three agencies in the business. She's managed to work her way up from the mail room to a desk job to Junior Agent in the literary department. Her

confidence is solid and her opinion is strong. She has all the savvy ingredi-
ents to make a powerful agent someday. She offered some of her thoughts
about why she's in this business and about the movies that may have helped
her make her decision to do so.

"Movies in general definitely have influenced my decision. Plenty of
people are in this business for the money (and there definitely is money to
be had) or the power or the abundance of free food at premieres, but some
of us really do love films. Go figure." She continues in her assertive, upbeat,
in-your-face manner:

> Believe it or not, I always loved *Swimming with Sharks,* and
> that one definitely has influenced me. I always thought the
> assistant needed to be tougher and to stand up to his boss.
> Granted, you need to figure things out at work to figure out
> exactly how to do this, but it can be done, and it can work to
> your advantage. Most agents, execs, et cetera, truly aren't evil
> and tyrannical, but there is much to be learned from the
> ones who are. And if you're in this business for more than
> five minutes, you'll definitely find them.
>
> A huge part of my job is figuring out what people will
> like. It's about going above and beyond what is trendy, and
> about setting the trends rather than following them. The
> bottom line is that fantastic material will rise to the top
> (though some crap ends up rising as well, but at least some-
> one is making a living), and when you see something that
> you're passionate about, you have to step up and not be
> afraid to champion it.

And champion is something that Heather has to do daily as she is up against
her competition consistently:

> Everyone is constantly looking for the next big thing, or
> looking to be a part of the current big thing. You really have
> to take care of clients in order to keep them around. If you
> don't, then someone else will. Clients are often loyal to their
> agents or agencies, but you have to nurture these loyalties,

and actively engage yourself in your client's interests. Rather than just throwing them onto one project after the next, you have to look down the line as to where you would like to see their career headed, and build the path for them. Not only is this more rewarding for you, but it keeps everyone happy, and if they're headed in the right direction, they don't mind so much that you're getting ten percent for it.

We're always looking for whatever that next thing, person, or idea may be, which is what keeps us going. The prospect of finding that talent and nurturing it to its fullest potential is fantastically exciting. And this business, at its best, certainly should be nothing short of fantastically exciting. If we can make people feel how they felt when they saw *Star Wars* for the first time, we're doing our jobs right.

CHAPTER 3

ASSISTANT

CREATIVE CAREERS IN HOLLYWOOD

STATUS

DURABILITY: Shredder (most definitely).

LENGTH OF STAY: One Year.

FOOD-CHAIN VALUE: Bottom-feeder.

UPWARD MOBILITY: Good (hopefully, as one can only move up from here).

DESIRABILITY FACTOR: High for those without ties to the industry, low for those who think they are part of some sort of privileged Hollywood royalty.

VACATION: Possibly, but the temp might take your job.

SALARY: Regular black coffee of the day.

HOW EASY IT IS TO GET THIS JOB: On a scale of 1 to 10 (1 being the easiest), 3.

PREREQUISITES: A pulse. If you are cute and dress well, that'll help. Drive and a good work ethic will get you through the humiliation. Leave your graduate degrees at the door.

I fear your young wards are little Antichrists in the making. You must tell them
that BMWs, car phones and Armani suits have a price that has nothing to do
with money. Not that they'll care. . . .
—Jeff Kurz, producer

More than any other industry, the entertainment industry
prides itself on the fact that you can be a high school dropout or a Ph.D.,
but you still have to pay your dues. In this business, most individuals
begin to pay those dues in their first jobs as assistants. Some are lucky and
have family ties in the business and are able to skyrocket into positions of
middle management and power immediately, although even they still have
to have the goods to continue in those positions. Most average Joes partake
in some sort of on-the-job-training situations when they first arrive in
Tinseltown. Those on-the-job-training situations are also sometimes known
as assistant jobs.

It is fairly easy to land an assistant job, and it is a well-known fact that
those who already know the area of work they want to be involved in (i.e.,
marketing, development, casting), are surely ahead of the game. If you don't
know what area of the Hollywood scene you want to specialize in, then sign
up with a temp agency and have them send you to various offices around
town. You'll get a quick week-by-week glimpse of what each department
and production facility looks like, and from there you can make more
informed choices. You have no idea how important those first contacts are.
Keep in mind that you are in the beginning stages of creating your contact

files. It is extremely important that you display your highest drive and deepest desires to succeed—even if you are only making $500 a week with no benefits and working until midnight every night.

THE FUNDAMENTAL FILM

There is really only one seminal film that addresses the role of Hollywood assistant, and that is *Swimming with Sharks*. Discussion of that film will follow, including lead characters, assistant Guy (Frank Whaley) and studio executive Buddy Ackerman (Kevin Spacey), but first, let's take a look at the varying degrees of assistant jobs and at how one goes about finding those assistant jobs in Hollywood.

ADMINISTRATIVE PERSONNEL

Finding your first job is actually much easier than any other job you'll ever find in your career. After a number of years, you will hopefully become an expert in one or two areas of the business and it will be difficult to break out of those niches. As a rookie just starting out, you can choose what you'd like to break into—and change your mind again and again if you want. Do it now, while you still have leverage and not that much invested. It'll get harder and harder with every year and every position.

Assistant

This individual is an executive's right-hand man or woman. Usually found within a studio, network, or production company, this person handles heavy call volumes; schedules meetings, lunches, screenings, travel arrangements, and events for supervisor; protects the executive in all ways and manners; keeps things running smoothly; and takes care of all administrative duties.

Executive Assistant

Generally, executive assistants function as basic assistants; they just do the same thing for an individual who is very high up in ranks, a vice president or CEO type. Often, the most powerful individuals have two or three assistants, one of which would be designated as their "first" assistant or executive assistant, with the second assistant and third assistant following in rank.

Personal Assistant

This term describes the assistant who caters to a single individual on a one-on-one basis, such as a celebrity's assistant. This type of an assistant, often referred to as a "handler," handles most of the celebrity or VIP's personal and professional life. Also performs all the duties of a basic assistant.

Receptionist

In a way, receptionists are assistants. They are assistants to the entire company and are located within the waiting or reception room of that organization. They also handle the heavy influx of calls and visitors and make sure things are running smoothly in the front office.

Mail-Room Person

Many successful Hollywood moguls started out in the mail room. It is an advantageous start to a brilliant career, for you have access to everyone in the company *and* you know the type of mail they receive. Not much pay here, but the experience is priceless.

Internships

Interns and gophers equal wannabes—wannabes who are doing something to get to where they *want to be.* Taking on an internship or being a production assistant (PA), a.k.a. gopher, are excellent ways to get behind the scenes and act out all of those fantasies. From those experiences, you make contacts and begin to work your way up the Hollywood Food Chain. Paid or unpaid, these are great opportunities.

BEFORE HOLLYWOOD

Film school or no film school? That is the question. And the answer is that education of any sort has very little to do with your success in this industry.

Earning a bachelor's or advanced degree is necessary for many jobs; however, it is not a requirement in Hollywood. Drive, determination, charisma, self-confidence, durability, and stamina often overrule education and intelligence in this industry. Therefore, it is not necessarily a prerequisite to have a degree from an Ivy League school or a second degree from any type of higher learning institution.

Having a bachelor's degree is an extremely desirable thing in today's workplace, and one should pursue the completion of that degree no matter what the chosen field of study. To begin working as an intern or assistant upon graduation is good. This is a hands-on industry where schmoozing is part of the game, and one can make more contacts working in the real world. If you want to be successful in this industry, go get a job, or go out and shoot your movie on your own, whether in or out of the school's curriculum. Be the writer that you are, be the director that you are, be the actor that you are. That's the quickest route to success. However, if you choose to go on to grad school to specialize in a film- or media-related path of study, make sure you make good contacts with your fellow students. Once you leave grad school, you're still going to be in the position of having to find a job to break into the industry, and that job could be as an assistant or alongside your fellow classmates as you all come together to form your new production company.

Before leaving college, make sure you have at least three examples of your best work—three screenplays, three short films, or three spec scripts (scripts written on speculation, not for the intent of any specific production company or producer, please see chapter 10 for further discussion of spec scripts). And, for aspiring actors, have a demo tape completed. Then, when you are working at a desk and one of the execs asks you what you are interested in doing when you grow up you can say, "Be a director—want to see some of my short films?" Be ready. Be bright and capable. Be prepared for when the windows and doors open. And remember that often the people you go to school with later become important, influential resources in the industry, especially if you go to a school that has a great alumni association that can help place you in your first job.

LOCATING A JOB LEAD

Finding a job in Hollywood can happen at any time during your professional and personal entertainment industry life. Combing the employment ads of the trades and the online job sites is mandatory, but you might also land a lead (and maybe the job itself), through word of mouth or by just being at the right place at the right time. Remember that showing up is 95 percent of the game. By just being in the Hollywood environment you may absorb information about an assistant who is leaving to go on to bigger or better things, or a job that may be created due to a new production company being given a first-look deal by a studio. By all means follow up on all leads—in print, online, and word of mouth heard over coffee or at the gym.

One example of an insider lead: get your hands on the UTA (that's short for United Talent Agency) Joblist. The list cannot be accessed on a Web site; you must know someone who already receives it. If you two are friendly, that person will hopefully forward it to your e-mail address. It is often a thirty-page-plus document that opens with a smattering of executive positions and immediately goes into assistant-level positions and lists of internships. This is just one of many Hollywood hotsheets.

One of the most honest of all assistant-level ads appeared in the November 11, 1998, edition of *Daily Variety* and went like this:

> ASSISTANT TO THE PRESIDENT
> Busy and exciting multimedia production company with offices in LA & NY seeks an Assistant with the following skills:
>
> • Windows, Word, QuickBooks and Computer Proficient
> • Able to work in an exciting but stressful environment
> • Detail-oriented with good organizational skills
> • Multi-tasking, accurate & make very few mistakes
> • Good people & communication skills
> • Even-tempered—positive attitude
> • Willing to travel

DO NOT APPLY if you:

- Want to be an actor or performer
- Are thin-skinned, too sensitive, feelings are hurt easily
- Can't work with temperamental creative types
- Want advancement other than this position
- Don't like working long hours
- Want a standard corporate structure
- Family & social life takes precedence over job
- Looking for short-term employment
- Don't possess the skills required

PLEASE APPLY if you:

- Want to be the top assistant to an entrepreneur who is fun, exciting & a mover & shaker
- You like being the top assistant & you are not looking for another type of job
- Are fast on your feet & can keep up with your boss
- Like challenges & changes & creative people
- Are solution-oriented, not problem-oriented
- Are willing to put in as many hours as it takes to get the job done
- Are not worried about family & social life
- Want to work for an established growing company with exciting projects
- Are looking for long-term employment

We would like to meet you if working for a company that rewards loyalty & dedication means something to you. Good salary & benefits. W. Hollywood office. Please fax resume.

All other online job sites, the trades, and personnel departments at all of the studios and networks are good places to start. Also, be sure you have on hand your *Hollywood Creative Directory,* which will list all of the players in town, and access to the Internet Movie DataBase (*www.imdb.com*). In addition, having the current editions of the *Videohound Golden Movie Retriever* and *Leonard Maltin's Movie & Video Guide* will help you in case you need to quickly look up a movie referred to within an interview.

CELLULOID ASSISTANTS

It is fairly certain that Guy, assistant extraordinaire and lead character of *Swimming with Sharks*, discussed below, has readily consulted these texts before his opening discussion concerning actress Shelley Winters—for whom he is able to list nearly all of her most famous credits through a fifty-plus-year career.

Swimming with Sharks (TriMark, 1994)

Swimming with Sharks is prerequisite viewing for anyone considering any type of work within the entertainment industry. The Buddy Ackerman–type executive exists across the board, whether he takes the form of an agent, a director, or even (most especially, actually) an actor. *Swimming with Sharks* revolves around the relationships between three main characters: assistant Guy; his boss, studio executive Buddy Ackerman; and an independent producer, Dawn (Michelle Forbes.) Most of the interaction takes place between Buddy and Guy, even though Guy and Dawn enjoy a love affair. Dawn, the neutral voice of the film, is heard in the opening monologue: "In Hollywood, one of the fastest ways to the top is to work for someone who is already there. The system dictates that one must first be a slave before you can become a success. For this can be a very demanding process—only a few people have the drive to endure the thousands of indignities and hardships that make up the system. This drive is usually motivated by greed, sometimes ambition, sometimes even love."

The newspaper ad for *Swimming with Sharks* stated "In Hollywood all his dreams could come true. . . . But first he has to make coffee." The words "backstabbing," "ruthless," "two-faced," and "revenge" are boldly placed behind a picture of Buddy yelling down at Guy. There was a time in Hollywood's history when filmmakers had to earn the right to bite the hand that fed them. In other words, many of the earlier films about Hollywood were written, directed, or produced by seasoned pros. George Huang was barely out of film school when he wrote *Swimming with Sharks.* He used his own experience as the basis for the film. Huang, who has been described as an upbeat personality who punctuates every other sentence with a ripple of laughter, interned at Lucasfilm and assisted several powerful executives during his five years as a lowly assistant.

The major relationship in this film, between a high-powered studio exec and his beleaguered assistant, unfolds in a series of flashbacks. Writer Huang has given the world a glimpse of what it is like to work under the incredible pressure of a maniacal boss. "Shut up, listen, and learn," is Buddy's mantra to Guy. To say he lacks tact is an understatement. Buddy is a tough boss. If Guy can live through Buddyland, Guy will be able to do anything he wants for the rest of his career. "Before you go out and change the world, you have to ask yourself, What do you really want?" is Buddy's advice to Guy. Not bad advice. Those who know what they want usually find themselves on a faster track to the golden ring, carrot, or Holy Grail than those who are still testing the waters. Guy doesn't know exactly want he wants. It hows as he basically allows Buddy to run all over him.

Perhaps the scene that tips the scales in this relationship is the famous "Sweet 'N Low scene." If you intend to be an assistant, you must watch this exchange—and watch it closely. Guy brings Buddy his coffee along with a sweetener. Guy, unaware that anything is wrong, gives Buddy his coffee and prepares to move on to other business. Buddy begins his tyrannical litany: "What I am concerned with is detail. I asked you to get me a packet of Sweet 'N Low. You bring me back Equal. That isn't what I asked for. That isn't what I wanted. That isn't what I needed, and that shit isn't going to work around here." Guy defends himself with "I just thought . . . ," to which Buddy screams at Guy that he has not been hired to think. (See chapter 9 for Buddy's side of this story.) These episodes of terror and torture go on as Buddy continually reminds Guy that he must earn his success to differentiate himself from his "MTV, microwave-dinner generation" who want everything without having to work and persevere.

A year passes. Guy is tired of Buddy's incessant demands and odd-hour calls, along with his deceitful behavior and empty promises. Buddy has even done the unthinkable: He's pitched one of Guy's projects to the studio as if it were his own. Such is the reality of Hollywood, and such is the truth that those who work hard become jaded in the process. In turn, it is highly probable that Guy will visit the same terror and torture upon others when he is finally in the power seat. This vicious circle continually feeds upon itself.

Buddy's constant berating leads Guy to his breaking point. He takes matters in his own hands, and, seeking revenge, takes Buddy hostage in Buddy's own home. Recalling what Buddy has put him through, Guy does his best to make sure Buddy is tortured physically, as he knows that not much can penetrate him emotionally. Finally, Buddy breaks down and reveals the truth about his wife's death. Says Buddy: "Your job is unfair to you? Grow up, way it goes. People use you? Life's unfair? Grow up, way it goes. Your girlfriend doesn't love you? Tough shit, way it goes. Your wife gets raped and shot, and they leave their unfinished beers . . . stinking longnecks just lying there on the. . . . So be it, way it goes. . . . "

Buddy's pain is expressed through the nastiness he unleashes daily. Everyone has his or her own personal hell to live through. Buddy's truth is revealed so the audience can understand Buddy's behavior and become sympathetic to this character's motivation. This doesn't mean Buddy's actions are justified, it just helps Guy and the audiences understand the reality of the character's makeup.

In most of *Swimming with Sharks*, writer Huang has given us a realistic look at the assistant's position. Linda Buzzell, author of *How to Make It in Hollywood*, preaches that there are only two sins in Hollywood: to be dull and to be desperate. Throughout this film Guy appears to be excruciatingly dull—and uniquely desperate, when he takes Buddy hostage. When Guy snaps, the film takes on a new energy, an energy of a mad-slasher film. It is as if Huang wanted to make the ultimate statement about cutthroat competition. But it goes without saying that in the real world Guy's action might not lead to the promotion he receives on screen.

DO YOU LOVE IT ENOUGH TO CHANGE IT?

Being an assistant is only the first hurdle to get through to prove that you are a player in this industry. You must ask yourself: "Do I love the industry enough to change it?" If the answer to that question is yes, then learning all you can as an assistant will help to propel you to the position of power you need to be in to make those changes. It all begins with that first job. Your good reputation will be generated and will follow you throughout your career. Shut up, listen, and learn. Good advice for the newly hired assistants.

IN REAL LIFE

An assistant is an assistant—whether he works for an executive or for a celebrity. Here are two competent assistant voices.

Assistant to a Celebrity

Nicolas Weststate is an assistant to a celebrity. When interviewed, he joked that maybe he should be referred to as an assistant to a *minor* celebrity, which might be more appropriate, for Nick's celebrity boss is a young actor who has had a few starring roles in major feature films and has not made $20 million a picture—yet.

Have you seen Swimming with Sharks?

The entire first year of my job I was completely afraid to watch *Swimming with Sharks.* One of the guys in my office (not my boss) was a complete Hollywood prick like Kevin Spacey's character—in fact I think he modeled himself partially after the character (he thought it was cool) because he always talked about how much he loved the movie. But any time it would be on TV and I'd be flipping through, my stomach would automatically tighten up and I'd get this sick feeling. And I hated the guy in my office for liking the movie, because I felt he did so for self-important reasons, like isn't it cool that my life mirrors a movie, rather than because it was a good movie. Now, fortunately, that era in my life has passed and I can watch it and laugh and say, as many do, boy, I remember having to deal with crap, just like Guy.

Is it more fun being an assistant to a celebrity than just to an ordinary person?

Generally, no, it's not more fun to be an assistant to a celebrity—doing assistant work is doing assistant work. However, at first, it felt important in some ways, because wherever I'd call, most everyone would know my boss, and give me exceptions and allowances. Plus, I'd often get my boss's extras—not only stuff like tickets to premieres, but also clothes, gadgets, great food. So, at first, I was a little wowed by the celebrity of it all, but after a little while, I became nonchalant.

What do you want to be when you grow up?

The career I am working toward is directing films. Definitely, this job is helping me get there. Not only has it been a great introduction and immersion into the industry, but also I have made worthwhile contacts to help me in my future endeavors.

What advice do you have for others who are about to enter the big world of working in the industry in L.A.?

Be open to a variety of experiences—just keep your eye on your goal (and be aware your goal may change) and everything else you do will be part of the journey toward that goal. Follow your instincts, but don't turn something away because you feel it doesn't directly relate to your goal—you can make it work. Acquire a variety of experiences—it will make you stronger at whatever you do. And be persistent. Don't give up. Your first year or more can really suck in a lot of ways, but you'll get through and be better because of it. Know also that you can't control or predict what will happen—accept that, roll with it, and learn to appreciate the good little moments.

Assistant to a Studio Exec

Patrick is an assistant to a studio executive on one of the major lots. He's paid his dues working for some of the more demanding and notorious individuals in the business and has settled in behind this desk over two years ago. Patrick is more than just an assistant; he feels he is a key player in the day-to-day operations of the studio system.

Have you seen Swimming with Sharks?

I haven't seen many movies about the movie business in their entirety. It's kind of like trying to read an insider book about the process of writing; it can feel a little intractable rather than enlightening if you don't walk in the same-sized shoes. I've seen *Swimming with Sharks,* the first part of *Living in Oblivion,* and part of *Bowfinger.* I did see *Swingers* just before I came to L.A. and it probably had the biggest impact on me. Not because it was tangentially related to the movie business but because it seemed to capture the pulse of the Los Angeles that I envisioned myself being a part of. Just as, say, *Menace to Society* conveyed the Los Angeles I wished to avoid.

Is it advantageous to be an assistant to a powerful person?

The best commentary I've ever heard on making it in Hollywood came from a successful producer who was describing a lecture he had given to Northwestern University students who were eager to start their careers. He advised them to stay home. Don't come. That none of them had any concept of how difficult the road to success was. He reasoned that most people could save themselves the trouble of finding out it wasn't for them by better spending their energy elsewhere. Of course, though it was fair warning, it wasn't going to stop anybody. It did however, crystallize something. Whatever you think you know or whatever strategy you devise to bolster your chances at "getting ahead" or "making it"—it's all going to go right out the window.

This is an industry that feeds off talent. Everyone has it; some can tap it better than others. Stick it out long enough here and you are going to have to face the facts about how much you have, in what ways you've spent your life developing it, how accessible it is to you, how far it is going to carry you, and whether this is how you want to spend it. There is no "path" to great success. Great success is forged in the frontiers that have no paths. You have to be visionary just to crack open those doors. You have to have the intelligence to carry yourself through. You have to have the singular dedication to see your vision past every obstacle. This is not hyperbole or apocryphal rambling. This is how you are going to feel. This is what's going to happen to you. If you plan on getting out of the minor leagues and playing in the big game, it's going to change who you are as a person.

Working at the top and bottom of the food chain both has benefits and detriments. I recommend doing a bit of both, but most important, take them for all their worth and know when it's time to make a leap of faith and get out. Stress? Deal with it. Realize that if you are lucky in life you will have to deal with such things as marriage, birth, and death—career should be small potatoes. It's a very tough business and you will learn when you put your foot down and stick up for yourself . . . a wall of fire of sorts. Allow yourself to get there, it may take six months or six years, but as long as you have been making the best use of your time while toiling away within the corporate hierarchy, you will not regret leaving. If you don't, however, stick with what it is you wanted to do, you will get lost in that rat race, your dream will die, and you will feel like failure. I've seen it happen too many times. Remember, that producer's speech could have saved you a lot of time and trouble.

CHAPTER 4

D-GIRL

CREATIVE CAREERS IN HOLLYWOOD

STATUS

DURABILITY: Shredder (disguised as a Keeper).

LENGTH OF STAY: Two Years.

FOOD-CHAIN VALUE: Mid-level.

UPWARD MOBILITY: Moderate to Good.

DESIRABILITY FACTOR: High, especially among assistants and wannabe producers.

VACATION: Too risky.

SALARY: Regular black coffee to Grande Latte (decaf).

HOW EASY IT IS TO GET THIS JOB: On a scale of 1 to 10 (1 being the easiest), 6–7.

PREREQUISITES: Knowing how to read. Being able to identify complete sentences. An extremely active imagination in order to envision scenes. Being really hot-looking.

I don't want to be a reader all my life, I want to write.
—*Betty Schaefer,* Sunset Boulevard

Working in development is a perfect example of a shredder disguised as a keeper. Development looks and feels like a keeper but it's a shredder. Being a d-girl is a great way to learn the biz, to cut your teeth, to learn the ropes, and all that, but beyond a two-year stint it's a revolving door to nowhere in the land of Hollywood milk and honey. The job has flexibility and, should you discover the right project (if someone higher than you doesn't falsely lay claim to that discovery), you could be propelled into a higher status, such as creative executive, producer, or vice president. This is a highly sought-after position within the Hollywood Food Chain, for it offers opportunity to those willing to grab it. Let's look at some d-girls who have been portrayed in the movies and at a little back story of the job.

DEVELOPMENT

Without scripts, there would be no Hollywood. And, in spite of what embittered screenwriters may think, without development, there would be no scripts. Development, for Hollywood, is the term used to refer to the unpredictable, circuitous, and frustrating process by which a script goes from the page to the screen.

Within this process, ideas are heard in pitch meetings, scripts and treatments (short synopses of storylines) are read, and notes are given as feedback to the writers and agents representing those writers. Also, elements, such as directors and actors who will be involved with the project, are attached. Development is really a birthing process. It is the time of preproduction that takes place for every project. Once the project is ready to go, fully edited and all elements attached, production, or birth, will commence, and those individuals in development will turn to new ideas to discover and nurture.

Within the lower echelon of every level of the Hollywood Food Chain, whether it's at a studio, network, or production company, you'll find people who work in development. That's it; not referred to as the "development department" but just "development." These individuals can be male or female. It is now considered derogatory to refer to them collectively as "d-girls" regardless of gender or age. Perhaps d-girls have earned a bit of a bad rap because the department they work in is a vast and undefinable platform with no real fixed boundaries. For the most part, d-girls are thresh-old guardians who are the first to yay or nay a script or project and are employed by studios and independent outfits run by producers, stars, and directors. They endure endless amounts of café lattes, lunches, and meetings in search of the perfectly producible script.

Anyone working in the following creative careers in Hollywood could be referred to as a d-girl.

Development Assistant

This entry-level position assists a development executive or the story department. The job is administrative. The assistant answers the exec's phone, handles her professional calendar (and also her personal life, including trips to the dry cleaner, BMW tune-ups, and weekends in the desert), and often prepares documents that track the number of projects the company has in development. There is some reading of scripts and submissions. This employee works in-house.

Story Analyst–Reader

This is also an entry-level position that can be freelance or staff. The reader provides coverage, a document that evaluates and synopsizes projects, to upper management.

Story Editor

This is a mid-level position. The story editor—head of the story department—supervises readers and becomes the point person that the staff and freelance readers pick up materials from and hand in coverage to. The story editor desk is the hub of the development department. This individual also reads scripts that are being considered for production, and may be called upon by upper management to do story notes, that is, give feedback on the story. The feedback will influence the suits into buying or passing on a project. This position is in-house, with occasional outside meetings.

Acquisitions

This mid-level employee is responsible for acquiring completed or nearly completed projects for the company. This entails in-house activities and fieldwork, such as attending film festivals and screenings of just-completed films that are looking for distribution or representation. An acquisitions manager will also hold meetings with writers and agents representing writers and directors, and wine and dine agents at luncheons and dinners to gain their attention and their A-list stable of clientele.

At the higher levels, there are the following d-girl creative careers in Hollywood.

Creative Execs (CE)

This is a highly sought after mid-level position. The individual usually works closely with a production executive or studio head, becoming the eyes and ears of the upper echelon, as the execs are too busy to plow through all the material being submitted. The fear of passing haphazardly on the next *Star Wars* puts the CE in a delicate place. The CE should read and evaluate every script or project he or she obtains. Creative execs are sometimes referred to as unsung heroes in the moviemaking process. These people work in-house and all about town. They know how to "work the room."

Vice President, Director, or Manager of Development

People in these three positions work with writers to improve a script. They oversee the production of the project with the vice president of creative affairs or the vice president of production, depending on the size of the company. These creative jobs are all in-house and all about town.

Creative execs and vice presidents may be referred to as d-girls, but only in a belittling or perhaps affectionate way, for these two positions are certainly priming for higher responsibilities. For the purposes of this chapter, however, let's look at and praise those lowly d-girls—the story analysts, readers, story editors, and acquisitions managers. These are the individuals who hate development hell as much as the writers and producers who must join them there. The creative executives and vice presidents will be featured at length in chapter 9.

WORKING BETWEEN THE REVOLVING DOORS

Development hell is the time it takes for the development department to

1) Read and evaluate any given script. The process of "waiting to get read" will be swifter if the script is repped by a big-time agency, such as CAA, Endeavor, or William Morris, or if it was just written by the writer of the blockbuster spec script of the summer.

2) Take meetings and agree to accept, option, or buy a script.

3) Get together, once a script is optioned or purchased, to read and make notes about the project to make it work for the department's purposes.

4) Get the entire department to read the script again, share notes, give the notes to the writer, and decide which direction the script should go.

D-girls, especially readers and story editors, can become like machines, sometimes reading five to ten scripts a weekend with twice that amount during the week. "I've actually begun to utilize a number of typical phrases describing usual story, character, and dialogue critiques as if they were part of one of those little boxes of refrigerator magnets. I just merely take a phrase and place it into my coverage—cutting and pasting phrases together over and over until the coverage is complete," says Eric, a veteran reader. D-girls churn out coverage like machines. They are under the professional gun to find projects that will please the production company's budget, the proposed actor's taste, and, last but not least, entertain the masses. D-girls are always in search of the next big movie—even if that's a small love story. They need to concern themselves with story, characters, and dialogue, and their quest for a story with a "hook"—the vague twist on a tried-and-true formula—will take over their lives. Coverage is a document that looks a great deal like a book report. There is usually a top sheet describing all of the particulars of the script—the author, number of pages, who is submitting it to the studio, and numerous other facts followed by a one- or two-page synopsis of the storyline. In addition to the cover sheet and synopsis, there will be a page of comments regarding what is working (good) and what is not working (bad) about the script. Coverage is a document that can make or break writers' careers.

D-girls scan solicited (submitted by a reputable agent or individual) and unsolicited (submitted by someone who is not represented by an agent, manager, or lawyer) material. D-girls can spot an amateur submission when a script is wrapped in plastic, spiral-bound, or has more than two gold brads within its three-holed paper. Armed with this criteria and continually asking the question "Why make this story now?" the d-girl begins her search in the trenches of Hollywood for the perfectly producible script.

D-GIRL BURNOUT

One can understand why d-girls sometimes crash and burn. With the lack of good scripts, that is, scripts that have actual coherent storylines, the journey becomes desperate, with everyone frantically searching for the same perfect projects. Everyone is competing for the same projects and no one really knows what those are, they just hope the script will have a boffo box-office opening weekend or soar in the ratings if it appears on the small screen.

THREE IMPORTANT D-GIRLS

There are three iconic versions of d-girls that have been portrayed in film. The following three movie characters provide a realistic perspective on what goes on in the trenches.

Queen: Betty Schaefer

All hail Betty Schaefer, the character played by actress Nancy Olson in 1950's *Sunset Boulevard.* As the character who said "I don't want to be a reader all my life, I want to write," Miss Betty Schaefer is the queen of all d-girls. The story of one Joe Gillis (William Holden), down-on-his-luck screenwriter who happens upon the Sunset Boulevard driveway owned by Norma Desmond (Gloria Swanson), is tragic. The ways and manners of Betty and her d-girl world weave themselves intricately into the plot of this classic Hollywood film.

When Betty is first seen in the movie, a production executive Mr. Sheldricke (Fred Clark) calls her to join him in his office. Betty's caramel-blond hair is tied back in a bun, her white shirt with Peter Pan collar is buttoned up to her chin, long-sleeved sweater and calf-length skirt completes the ensemble, with a bow in hair and Max Factor–style makeup. She is a picture-perfect example of a mid-twentieth-century working girl. She is asked for a copy of her coverage and her opinion of a story she has just read. She responds quickly, confidently: "It's a rehash of something that wasn't very good to begin with. I found it flat and trite." Mr. Sheldricke then introduces her to Joe, the author of the "flat and trite" story. She is taken aback, makes a gesture to be polite, but defends her thoughts by stating: "I just think a picture should say a little something." The three discuss how writers could take Plot No. 27A and make it glossy, make it slick, until Joe states he needs to write to make a living—and, basically, it's coverage like the one she's just provided that's standing in his way. True, Betty did wield that little bit of power by expressing her educated opinions in her coverage, but the meeting provides her with an introduction to what she begins to dream is her future. Fate, of course, has other plans.

Betty is an excellent composite of many young women who worked in Hollywood during its early years. She's fastidious, confident, and determined. The next time she sees Joe, she's at Schwab's Pharmacy with her fiancé Artie (portrayed by a young Jack Webb, a.k.a. *Dragnet*'s Sgt. Joe Friday). It's New Year's Eve, her hair is still combed back, but she's ditched the white blouse for a strapless (yet demure) dress. She tells Joe that she felt a little guilty about their first encounter. As a result of that meeting, she took a second look at some of his old stories. They begin by discussing business but the interaction turns into a delightful exercise in character dialogue. Joe says, as he leaves to return to Norma, "You'll be waiting for me?" Betty replies: ". . . with a wildly beating heart . . ."

And that she has for him. Betty is then seen fast at work at her desk on the Paramount lot. With pencil in hand, she dials Crestview-51733, desperate to locate Joe. She is told Joe is not at that number. It's back at Schwab's that she sees Joe again and again, but she's with Artie (the famous drugstore is their favorite hangout.) She explains that she's been calling him. Joe announces that he's given up writing on spec—in fact, he's given up writing all together. It is here that Betty, disappointed, announces that she wanted to get in on a deal with him—she had twenty pages of notes on one of his stories—and she delivers her famous line: "I don't want to be a reader all my life, I want to write." Apologetically, Joe says, "I'm sorry if I crossed you up."

"You sure have!" she exclaims.

With Betty's determination, she won't give up. Joe sees Betty on the Paramount lot. She tells him that Artie is on location in Arizona, working as an assistant director. She's free evenings and weekends. Betty and Joe throw around a few ideas, he still tells her "no" playfully, as he leaves and she nearly tosses her half-eaten apple his way.

The next shot is of Joe, standing in Betty's cubbyhole of an office, telling her that he feels like he's playing hooky every time he escapes from Norma's mansion to come and work with her. Betty makes coffee in her sweater set and long A-line skirt. They smoke cigarettes as they type away page after page. They take a break and tour the back lot, apples in hand, a picture of innocence. Betty reveals that she was born two blocks from the studio—her father was an electrician, her mother still works in Wardrobe. She's had ten years of diction and dancing and a new $300 nose after a studio test. The studio didn't like her acting so she gave it up—"it taught me a little sense"—as she worked her way from the mail room to stenography up to reader. (See? Even then the Hollywood Food Chain was in full force.) "What's wrong with being on the other side of the camera?" she asks. Joe's cheering for Betty as things heat up between them—a near kiss—he tells her she's like "freshly laundered linen handkerchiefs" and asks her to stay two feet away from him at all times. A few scenes pass . . . and finally Betty admits to Joe that she's no longer in love with Artie. Joe asks, "What happened?" Betty answers, "You did." They kiss.

Meanwhile, back at the palatial mansion, Norma, who has now taken possession of Joe, discovers a cover page from a script Joe is writing, *Untitled Love Story.* She doesn't like this. Her jealousy is driving her into frenzy. She's already attempted suicide. Norma places a call to Betty's home, which results in Betty and her roommate Connie being invited to visit 10086 Sunset Boulevard—Norma's abode. They arrive only to find the odd menage of has-been Norma, her creepy manservant, and the very surprised Joe. Betty, a trooper to the end, states that she never received a call from Norma, she's never been to Norma's house, and, wanting to erase the weird visit, begs Joe to go with her. He tells her no, wishes her luck, and tells her to finish writing the script while on her way to Arizona to return to Artie . . . and that's the end of Betty's character in this all too tragic Hollywood story.

Betty survives the ordeal. The picture of confidence, this twenty-two-year-old kid with plans to be a writer more than probably succeeded in a writing career. Betty Schaefer is the first and original d-girl.

Princess: Bonnie Sherow

While Betty was there, turning on a dime, and Mr. Sheldricke called for her and her coverage, it wasn't Mr. Sheldricke that Betty fell for romantically—Betty fell in love with the writer, in her creative career in Hollywood. In *The Player,* d-girl Bonnie Sherow (Cynthia Stevenson) falls for her boss, Griffin Mill (Tim Robbins), the modern-day equivalent of Mr. Sheldricke. With the decades that pass and the changes in women's roles, the job of d-girl takes on certain ballsiness. Peter Pan–collared white blouses and long skirts have become tight-fitting, pastel-colored, mini-skirted power suits, with a gold watch and gold earrings and black high-heeled pumps as accessories. Bonnie's opening scene features her scolding her assistant Whitney Gersh, portrayed by actress Gina Gershon (who would in a few years surpass Cynthia Stevenson on the popularity scale) by stating: "You're my assistant, you don't get involved with writers!" (So much for keeping the development assistant on the low end of the Hollywood Food Chain.) Bonnie enters Griffin's office and jumps into the lap of her boss (also her lover). She asks if they can go to the Springs (Palm Springs) for the weekend—as she is in desperate need of a massage and a long soak in a hot tub, with margaritas to be administered intravenously.

Later, at a Hollywood party peppered with many famous faces, Bonnie goes gaga over Harry Belafonte and, when introduced to Marlee Matlin, begins to discuss script changes. Griffin jumps in to interrupt her, explaining that she should never talk shop at a party. (He's right; parties are for that great art of schmoozing. Flatter and bat the eyelashes, but don't talk shop until the following Monday, when you make that phone call to persuade that hot young writer to rewrite your latest project-in-development.) Nonetheless, Bonnie's big scene follows when she and Griffin are in his hot tub. Bonnie reads a ridiculously steamy sex scene from a movie script as she reveals her tits. She ends the scene straddling and kissing Griffin, saying, "Can we go to bed now? I'm starting to wrinkle." This is a far cry from Joe's comment to Betty about her smelling like "freshly laundered linen handkerchiefs."

Back at work, in fact, back in the screening room, Bonnie watches dailies with Griffin and a number of other studio regulars. There's a reference to Joe Gillis when someone calls to make an appointment with Griffin, identifying himself as "Mr. Gillis." The crowd refreshes Griffin's memory, reminding him that Gillis is the murdered writer in *Sunset Boulevard*. The line foreshadows the plot of *The Player*.

In another scene, Bonnie steadfastly stands up for Griffin when he is in danger of losing his job. Bonnie does what any good d-girl should do—she defends her boss. Griffin, now no longer interested in Bonnie sexually, sends her to New York to close a deal with a writer. Gathering her things to take the trip to New York, including ten scripts, she is flustered, for she knows something is very wrong with her romantic relationship with Griffin. (He's met someone else—the dead writer's wife, actually.) He tells her to go to New York because if she scores there, the studio head will make her a VP.

"Are you afraid of success?" he asks her. No, she's not afraid of success, she's afraid of losing him. Betty wanted a romance and a cowriting career from the man she loved; however, she was never able to bring the relationship to fruition. Bonnie is also denied a romance by her unfaithful boss. In both cases, these d-girls wanted romance and successful careers, but their careers managed to survive longer than their relationships. So again, the d-girl continues on her career path.

A year passes. Bonnie has been working the revolving doors of development—single and bitter. Her views are no longer respected, what with the new regiment of executives and the changeover at the studio. At a screening, she expresses her unpopular opinion regarding the studio's upcoming favored release. She is fired on the spot. Bonnie, in pure nineties d-girl fashion screams back "Fuck you!" and loses a heel off of her pump while walking back to the office. Her assistant, Whitney, is seen whispering into the new exec's ear—looks like she'll be moving up that Hollywood Food Chain ladder any day now. Her talking to the studio execs and the writers, which Bonnie warned her not to do, has most definitely paid off.

Poor Bonnie—she is passed over by Griffin as he walks out of his office, refusing to meet with her. He tells her, "Bonnie, you'll land on your feet, I know it." She sits on the steps of the office building, barefoot and crying. Exit d-girl Bonnie Sherow. The many reviews of *The Player* often refer to Bonnie's integrity. It's as if she is the only one of the players who is strong enough to stand up for her own opinions, her own point of view—similar to Betty Schaefer in focus and scope.

Bonnie is the princess of all d-girls. She does her best to defend her scum-sucking boss, and is later slapped in the face and fired for her righteousness. Telling the truth—your truth—is not always the right way to go if you want to keep your current creative career in Hollywood. Especially at the lower end of the Hollywood Food Chain.

Betty versus Bonnie

Few things have changed in Hollywood in over forty years for here are two women, decades apart, who experience nearly the same trials and tribulations within approximately the same job equivalent. Here are some comparisons between the Queen and the Princess.

	Betty	Bonnie
Work Outfit	White shirts, calf-length skirts, with black pumps, sensible watch, no earrings. Max Factor make-up.	California-color power suits short skirts, black pumps watch and earrings.
Previous Employment	Paramount Mailroom, Steno Department. Did a screen test but failed. Destined to work behind the camera, not in front of it.	"Used to be at Tri-Star."
Starstruck	No. Hates the fact that the man she's fallen in love with lives with an old silent film star.	Yes. Goes gaga over Harry Belafonte at a party.
Likes to talk 'Shop Talk' at party?	Yes.	Yes.
Involved with boss?	No. He doesn't even know her real name.	Yes. They suck face and share hot tubs.
Involved with talent?	Yes. Betty falls for writer Joe.	No. She's got her boss.
Reveals herself	Yes. Confesses she's paid $300 for a new nose.	Yes. Shows tits in hot tub. Confesses she's getting wrinkly.
Favorite Snack	Apples	Kissing Griffin.
Coverage Comments	"I found it flat and trite."	"The lead is a fifty-year-old circus performer."
Road to Success, a.k.a. Hollywood outcome	Coscript an Untitled Love Story with Joe—it will get her out of the Readers Department.	Fly to New York to secure rights to a Tom Wolfe book—she'll get VP stripes.
Actual Outcome, a.k.a. Reality Based Outcome	Begs Joe to leave Norma, gets her heart broken and remains in Readers Department	Challenges the phony ending for potential block buster, fired, screams "Fuck you!" to her boss and cries.

	Betty	Bonnie
Smokes	Yes.	No.
Drinks	New Year's Punch.	Martini.
Last words from the Men they loved:	"You can finish that script on the way to Arizona."	"Bonnie, you'll land on your feet. I know it."
Famous Line:	"I don't want to be a reader all my life, I want to write."	"You're my assistant, don't get involved with writers."
Quote of Integrity:	"I just think a picture should say a little something."	"You sold out. What about truth? What about reality?"
Pivotal to Plot—the Truth Be Told	Betty discovers the truth about Joe being kept by scary old Norma.	Bonnie is the only character willing to speak the truth about how bad the product really is.
Earned D-girl Title	Queen, the true Original D-girl.	Princess. She does the D-Girl title right.

Miss Congeniality: Cathy Breslow

If Betty is queen and Bonnie is princess, then Cathy Breslow, played by actress Joely Richardson in James L. Brooks's 1994 *I'll Do Anything*, could be named Miss Congeniality within the d-girl pageant of stars. Originally a musical, this comedy features Nick Nolte as Matt Hobbs, a down-on-his-luck mediocre TV actor who is aging and has seen better days. When Matt's six-year-old daughter Jeannie is suddenly thrust into his life—due to his wife's decision to drop out of society—he is forced to grow up himself. While Jeannie auditions her way across Hollywood, Matt meets a beautiful movie executive, Cathy, who, after getting to know Matt, believes in him, or at least is hot enough for him sexually to give his career a jump start. Cathy's love and Jeannie's newfound stardom (that of "adorable daughter becomes a success") change the self-centered has-been actor into a mature and relatively happy adult father.

Cathy proves to be a perfect example of a revolving-door d-girl. She's a former acting student, now glorified script reader. She'd be terrific at evaluating screenplays if she had any idea what her own opinions were. She's lovely, but weak and lacking the courage to support convictions and ideas.

In real-life Hollywood, there are many d-girls like her. They feel compelled to side with their bosses, leaving themselves very little room to speak their own truths. Also, no one likes to make decisions in Hollywood because of the failure level—it's way too risky. Cathy's idea of a grand new movie is to do a remake of *Mr. Deeds Goes to Town*, yet she is the most non-Capraesque type of person that could ever exist. It is a bit surprising she even knows who Frank Capra is, actually; the general ignorance of baby d-girls is notorious. Cathy and Matt's relationship is typical—the d-girl attempts to promote the writer's, or actor's, or producer's career by developing a project that would be a perfect vehicle for him. This is an extremely convenient arrangement. Once again, like Betty, Cathy is involved with the talent.

I'll Do Anything, though no great box-office success, is actually not a bad movie—especially if you want to learn more about the entertainment business. Creator James L. Brooks is an industry veteran and a master at revealing what goes on behind the scenes. Perhaps the best scene in this movie is the gathering of creative execs, all Cathy's peers. The truth is heard here. Young, just barely out-of-school executives tear apart Tommy Lee Jones with his "very unfortunate skin," and a number of other leading men who are going bald. Matt defends the working actor by interrupting the meeting and explaining that it is the acting that is important—it's not all looks.

The scene is painfully close to the truth: this *is* how development execs judge talent—whether it is an actor, director, or writer. You are only as good as your last picture or only as good as your last blockbuster or, worse yet, only as good as how you looked at age twenty-five. If you fall below that level of excellence, your worth diminishes on the Hollywood Food Chain. At any rate, Cathy leads this group as an excellent example of a clueless-but-trying-really-hard-to-get-it d-girl. She would surely earn an honorable mention medal at least.

IN REAL LIFE

Some d-girls are famous—not necessarily for being d-girls, we shall see—and others are just really hard workers. D-girls are the worker bees of the development system.

Famous D-girls in History

There are very few resources to turn to to find information on the assembly line–like work situations of Hollywood's early days. The reading and typing pools were filled with script girls like Betty Schaefer, and every once in a while, a lucky one escaped the pool and was allowed to write. One of the first of these women was Kate Corbaley, a script reader for Irving Thalberg and studio head Louis B. Mayer in the twenties and thirties. She would read storylines and spin yarns for her bosses. She remained anonymous, due to the fact that she was a woman.

Kate wasn't alone. One of America's greatest writers and philosophers, Ayn Rand, was also a freelance reader. While writing short scripts and making suggestions whenever Cecil B. DeMille asked her opinion, Ayn was employed as a reader—first at RKO and later at MGM. The time was the mid-thirties. Her work consisted of reading books and manuscripts submitted to the studios, synopsizing them, and evaluating their screen potentiality. Solid d-girl work. RKO paid two dollars for a brief synopsis, and five dollars for a long one. She lived on that money, albeit modestly, while plotting and outlining *The Fountainhead*. She most certainly fulfilled Betty Schaefer's declaration of not wanting to be a reader all of her life.

In addition, Vicki Baum, author of *Grand Hotel*, worked as a scenarist and sometime script reader while in exile in L.A. from Germany in the thirties and forties. And famous diarist Anais Nin was also employed by numerous independent producers and directors to give her opinions regarding the potentiality of projects. D-girls are traditionally in good company. In the first half of the twentieth century, writers themselves or producers working alongside studio heads performed d-girl job duties. There were fewer ideas to process and attentions were only paid to a few projects at a time. The power to recommend or pass on a script lay at the top alone. It was not until the second half of the century that a department known as development was formed.

David the Superior Reader

"I'm a superior reader," David announces. David, late thirties, has been a professional reader for nearly ten years.

Reading is my main source of income. I have no plans to move up in the business. Only one in a thousand readers gets promoted and, let's face it, I'm not going to be one of them. In an average week, I read fifteen scripts, four or five at a time, for a major studio. I also occasionally read for a national cable network and a B-movie star's production company. From these ancillary resources, I usually receive an additional three or four books a week. At approximately $50 a script and $100 a book, I make right around $1,000 a week. It takes me one and a half hours to read and do coverage on a script, slightly longer for books. This takes up about forty-five hours a week and it is a draining experience. Reading consumes me. I find that I am often in dire need of downtime, brain drain. It might be the way I do coverage that is so exhausting—I read a script and type up a report, very rarely will I read three scripts and do three coverage reports in a row.

David is pensive, thoughtful, as he reports on his way of making a living.

I take extensive, copious notes utilizing symbols such as a heart to illustrate two characters that are, may, or will be lovers or "ma," "pa," "bro," and "sis," as abbreviations for family members. I use a great deal of "!" and "?'" as notes are taken. This is shorthand I've developed to help me read and comprehend the story as I go along. I don't keep the notes. The upside of this job is that I don't have to answer to anyone. I just have to show up to pick up the material, do the report, and return the material. Throughout the year, at any given time, I've read four scripts in the current Top Ten grossing movies—nearly a year and a half to two years prior to the movie's opening week. I know that I have some power to recommend or pass on a script but I don't feel like I've hurt any writer's careers, because ultimately it is the deal, the people handling the deal, and the material that makes or breaks a writer.

When he is asked what his secret of success is as a superior reader, his answer is:

> I capture the essence of the writer's work within my synopsis. I recreate the writer's style within the coverage. The synopsis has to be logical, a little story onto itself, and thoroughly readable. Overall, I have a Depression-era mentality. I take as many scripts and books as is humanly possible each week, because next week there may not be any material to read, hence, no money coming in. Reading is factory work. It's just like working on an assembly line.

Yes, David is a superior reader.

The Quotable Kelley

Kelley, twenty-five, is an assistant to a motion-picture literary agent at a major agency.

As part of her job, she is asked to read all of her agent's scripts. She acts as his "eyes and ears" and must provide him with full coverage of any projects she is asked to read. On the average, she reads five scripts a week, this is in addition to her regular assistant duties at an international desk. Her hours are 7:30 A.M. to 8 P.M., Pacific Standard Time. Often the scripts she is asked to read must be read and full coverage completed by the next morning. She can discuss the script directly with the agent and ultimately, the turnaround time for the script is quick. Many in her position move from the agency to a studio, production company, or casting office. As far as climbing up the Hollywood Food Chain utilizing her reading skills, she would only consider a CE position, allowing her to interact with executives, talent, and agents. She believes that any assistant who can do coverage — good coverage, that is — is valuable. It is not what the assistant thinks about the script, it is how the information is communicated. The script must be summarized in three pages, not twenty, and the coverage must make use of proper grammar and proper spelling. She will adamantly proclaim facts, such as the following: "This is a business. Baz Luhrmann is an artist (*Moulin Rouge!* — opening weekend $14 million). Michael Bay (*Pearl Harbor*—opening weekend $75 million) is a businessman." Does she have power? Yes. Her coverage stays in the agency's file for ten years—her recommendation is very important to the writer. On the other hand, her pass could equal demise to the writer. She could, however, find a job at a studio and remember that same particular writer and call upon him and his script many years later, making it a box-office success in its own due time.

The D-girl Dot.com Escapade

Ryan, twenty-seven, a writer-comedian, was referred to the new dot.com Web site via a very reputable Hollywood production company. Ryan had been its #1 reader for the last six months. Ryan was hired as a reader for the Web site that would post coverage on the 'Net for all of the Hollywood community to see. He was told to use a pseudonym when writing coverage, due to the fact that the information would be available via the Internet. In the past, when coverage had been written, the information found within the report would be kept private, only to be read by in-house employees. With the new technology, a script could get coverage and receive a yay or nay for the entire Hollywood community to see. Ryan refused the anonymity and went with his real name. He ran the risk of ruining his career as a reader, if he passed on a particular project and offended a portion of agents, writers, or producers in the community. The act of doing coverage privately had never before been challenged in this very public arena.

In the fall of 1999, on the first day the new dot.com was launched, fifty thousand people logged on to read the coverage of the scripts that had been read the night before. The availability of such sensitive information via the 'Net proved to be a disaster. One of Ryan's first scripts warranted a pass. He stated that it was the type of script Showtime would air at 3 A.M. Agents began to threaten a boycott of the Web site. Ryan continued with coverage; after all, he was being paid $100 a script. His next online entry had the opposite effect. This time, Ryan recommended an action-adventure female-buddy movie script, and within minutes, Tinseltown was a-buzz with the writer's name. She sold the script for millions and it made her career.

This action proved that via the 'Net, the reader had much more influence than ever before. Still, agents were furious, threatening to have the Web site shut down. Ryan was told to curb his comments. He refused to be censored and stood by his thoughts—after all, wasn't that the job he was hired to do?

Within a week, the Web site was scrapped. The dot.com caved in to the agents' request to stop telling the truth about their less-than-mediocre scripts and writers. Ryan returned to his pursuit of acting, having learned that one random kid's opinion could bust through the Hollywood bureaucracy. He had had enough of the bullshit involved with reading. In retrospect, he feels he learned a great deal and now has a commanding knowledge of story and can sum up a project in a perfect log-line in two seconds. He'll do coverage for friends but never again does he wish to pursue any of the positions in development. He's currently auditioning his latest folk opera at a Melrose Avenue cabaret.

CHAPTER 5

DIRECTOR

CREATIVE CAREERS IN HOLLYWOOD

STATUS

DURABILITY: Keeper.

LENGTH OF STAY: Lifetime.

FOOD-CHAIN VALUE: Very high.

UPWARD MOBILITY: You are very high—only heaven would be the next stop.

DESIRABILITY FACTOR: Tres high.

VACATION: Yes. Wherever you would like to go.

SALARY: Depending on the budget, black coffee to Grande Cap.

HOW EASY IS IT TO GET THIS JOB: On a scale of 1 to 10 (1 being the easiest), 10 (1 if you are willing to finance your own projects).

PREREQUISITES: Knowing the great directors and their work. Being able to control a movie set as if you were a godlike figure. Having a cinematography background. Knowing how to edit your "masterpiece."

Ed, visions are worth fighting for. Why spend your life making someone else's dreams?
—Orson Welles in conversation with Ed Wood in Ed Wood

How many times has the catchphrase "But what I really want to do is direct" been said by industry wannabes, young and old? Student film-makers tend to commit to this phrase as their mantra for life. Others get wrapped up in their "day jobs" within the industry and never really get back to their dream of being a director. Still others make their independent films and get them out to film festivals and agencies and find success, or wallow in indie-land never heard from again, or wind up directing commercials for advertising agencies.

A director has passion for making movies, movies of his choice. Being a director is a job that comes from within. It is difficult to understand the Hollywood of old, which assigned directors to projects within the studio system. Today directors generally make their first films and, if successful, are guaranteed the opportunities to continue to direct films of their choosing or they are attached to major assignments on projects that will be in the industry and public spotlight.

Being a director is like being God on the set. The director is the Supreme Being when a movie is in production. It is up to him to bring the vision to the screen, bring the words found within the blueprint of the script to the construction site where the movie will be built. Directors need to be creative, whimsical, and determined. Their authority must be consistently upheld on the set in order for the film to get made. This is an extremely popular creative career in Hollywood, and there have been numerous films made that have addressed the director character. In this chapter, we will look at seventeen films that celebrate this great job, but first, let's break down the types of directors generally found on the set.

DIRECTORS: THE DIFFERENT BREEDS

Depending on the type of production and budget, the duties of directors are broken down as follows.

Director

One of the highest positions on the Hollywood Food Chain. The director controls the set, telling the cameramen where to put the camera and the actors how to interpret the scene. Either you direct your own script or find a script you believe in; there are really no other ways to learn how to be a director. In order to gain recognition as a director you probably have to direct a short or feature-length film or else prove yourself in other media arenas, such as music videos, commercials, or even professional photo shoots.

First Assistant Director

Not always what the title suggests, assisting the director of the project. This is more of a technical position, a hands-on job that has to do with the functions of the production's location, cast, and crew. Essentially, this is similar to a line producer's position.

Second Assistant Director

Basically a coordinator of the project, a liaison between the First Assistant Director and the production office and other talent.

Second Unit Director

This director shoots all outside and peripheral sequences that don't involve the lead actors.

HOW TO GET STARTED

Directors vary in their expertise and the way they approach their subject matter. Some pay more attention to the details of the shoot and are known as cinematic directors. They carry on the traditions of the early Hollywood directors and the auteurs of foreign cinema. Such a director sees the film in terms of story and dialogue, placed within the framework of visual effects. Another type of director is the actors' director, who relies heavily on the actors and lets them take the story and dialogue where they will.

For those of you who like to take multiple-choice tests and would like to be a director, the Directors Guild of America, the DGA, offers a yearly opportunity to be chosen as an intern for the DGA. Their DGA Internship Test is available for taking at certain times of the year throughout the country. Even if you are not interested in being a director, it is a fascinating test to take. Contact the DGA in both Los Angeles and New York for details.

How to get started as a director? If you are lucky enough, you might be asked to take the test and become part of the DGA Trainee program. Another way to break in is to work as a "shadow," someone that follows the director throughout the shoot, learning firsthand how to do the job of the director. For the rest of you, the best way to become a director is to pick up a camera and start shooting a movie. Let's take a look at the many characters portrayed in films who have chosen directing as a way of life, or a way to tell their stories.

THE EARLY DIRECTORS

Films about filmmakers and directors are popular fodder for movies about movies. Perhaps it is because the director is at the center of the entire operation and knows this Hollywood life so well. From the thirties to the fifties, here are some of the first director yarns.

Sullivan's Travels (Paramount, 1941)

Preston Sturges decided to make a statement. Tired of the entire Depression-era social-issues cinema, he chose as his lead a director, one

Joe Sullivan, portrayed by Joel McCrea, to make his point. Sullivan has been producing comedies—successful comedies. Sullivan, however, aspires to be a "relevant" filmmaker, the kind that prefers to educate rather than entertain. He would like to make a socially significant film, such as the *Grapes of Wrath*, but his studio bosses try to talk him out of it. Their determination to keep Sullivan pigeonholed in the arena that has proven successful for him only inspires him to make his own plans reality.

Sullivan, a privileged studio director, is shielded from the hardship of the rest of the world's depression. He knows nothing of the poverty in America, which he proposes to portray in his art, so, in order to find out what it is like to be an average citizen, Sullivan leaves his cushy studio surroundings and takes to the road masquerading as a hobo. Through a series of adventures that include meeting the beautiful Veronica Lake, whose character is referred to as The Girl, Sullivan mingles with his public and lives among his audience and eventually learns humility. Sullivan learns what the common man needs. One of the basic needs is a release from his reality, and that includes humor. And humor, along with laughter, can be found in comedies. Thus Sullivan learns that he is, in fact, providing the world with socially significant productions, and he is able to return to his ordinary world knowing his work is good.

This Sturges's masterpiece is a comment about film in the thirties; however, the example Sullivan makes is one that could be followed to this day. If studio executives are holed up in their ivory towers at the studio, living their six- and seven-figure lives, sheltered from the realities of their public, how can they develop new properties that the average Joe will relate to? This is a lesson that everyone working in the industry should be aware of. To those of you who are about to embark on a career producing product for the masses, remember: don't forget your average audience, don't get caught up in a lifestyle that takes you away from the people who love the movies, for they are your bread and butter. *Sullivan's Travels* is a film that holds up no matter what century you watch it in.

Inserts (United Artists, 1976)

This 1976 Richard Dreyfuss movie focuses upon a one-time child star turned director, known as the Boy Wonder, who is now nearing middle age. Washed up after the talkies hit in the thirties, Boy Wonder has taken to drink and chooses to remain holed up in his palatial Hollywood Hills mansion. To make a living, he directs porn, welcoming to show business all the young starlets he can find.

This movie borders on a porn movie itself. The Depression is in full swing, and Boy Wonder doesn't have many options. Like director Sullivan, Boy Wonder reaches a point of desperation due to the social conditions he finds himself in. Boy Wonder knows there is little else he can do in the industry, and gets caught up in the seamy side of the industry. Sullivan explores his options and moves forward to enlightenment, following his adventures on the road. Boy Wonder's downward spiral is more frightening.

This movie received an X rating, pronouncing it dead on arrival at the box office. Dreyfuss had just experienced the mega-success of *Jaws*, and it was assumed that any film starring Dreyfuss would be a hit—until a few critics saw this raw little film, really just a study of a down-on-his-luck, out-of-work director, and panned it. The title refers not only to the obvious sexual reference but also to the shots that are inserted into a film, usually after the main portion of the film has been completed.

And so, this five-character, one-set film that utilizes color and black-and-white photography so well (color for the film the audience is watching, black-and-white for the movies being shot) works effectively as a snapshot of the time. In both examples so far, these directors are victims of their time. How can a creative job be so oppressive? The forties will prove to be more illuminating.

Hollywood Story (Universal, 1951)

Written by Frederick Kohner, the same writer who penned the fun-in-the-sun classic *Gidget, Hollywood Story* was narrated by a young Jim "Mr. Howell" Backus; directed by one of the best B-movie directors of all time,

William Castle; and starred Richard Conte as a young producer-director who arrives in Hollywood determined to make a movie about a mysterious murder that was committed at an old studio bungalow over twenty years earlier.

Larry O'Brien (Conte) is advised by his agent, his screenwriter, his financial backer (played by Fred Clark who appears in the same type of role in both *Sunset Boulevard* and *Dreamboat*), and the daughter of the silent-film star who was to appear in the tainted movie that never got made twenty years ago, not to do this picture. Undaunted, Larry pushes forward and "walks in on an old hunk of Hollywood history." This movie shows how a director can become completely obsessed with his work. O'Brien turns sleuth, uncovering, one by one, the clues to this unsolved murder. Throughout this journey, there is no directing work, nothing else is important in his life. His passion engulfs him, and because of the desire to tell the story of the death of director Franklin Ferrara, he solves the case. The result of his detective work and the uncovering of the murder is that he now has a script to shoot.

Castle captures the energy of Hollywood in all its innocence of the fifties. The production values are decidedly low-budget, but this movie does provide solid entertainment.

The Barefoot Contessa (MGM/UA, 1954)

The Barefoot Contessa opens with the funeral of a beautiful young star. This tragic drama is told in flashback from the point of view of down-on-his-luck director Harry Dawes (portrayed by Humphrey Bogart in one of three Bogart movies about the movies — see also *In a Lonely Place*, wherein Bogart plays a writer, and *Stand-in*, which has Bogart working as a producer). Dawes has been sent on assignment, along with the studio's press agent, to scout for a leading lady. They find her in the beautiful and captivating Mara Vargas (Ava Gardner), a dancer in a cheap bar in Madrid. Vargas plays hard to get, but Dawes follows her home to convince her that he must direct her in his next movie.

Vargas, who is always running around barefoot—hence the title—must cut her family ties to escape from her overbearing mother. She finds a way, and Dawes falls longingly in love with her. Dawes's movie becomes a smash hit, and Vargas is catapulted into the limelight, marrying an Italian count, which ultimately leads to her murder.

Dawes is the lead in this movie, even though Vargas plays center stage. It is his cynical narration that reveals the seedier side of the industry; like other movies about movies in the fifties, this one reveals the dark side of Tinseltown. It is Bogart's point of view as a seasoned industry vet who has seen it all, the ups and downs, the triumphs and the misery, and who himself is facing old age without continued success, that gives this movie a dark tone. His love for this beautiful woman, perhaps his last chance at love, ends in tragedy, adding to the element of darkness that this movie portrays.

Nonetheless, this movie is a must-see for Bogart fans as his usual odd and edgy intensity fills up the screen. His confident portrayal of a movie director, one that has seen it all—now including death—is solid, and it is a portrayal of a director's life well lived.

Ed Wood (Touchstone, 1994)

For another good Hollywood story from this same era, look no farther than *Ed Wood*. As far as motion-picture history goes, Ed Wood is regarded as one of the least talented movie directors of all time. Tim Burton's 1994 *Ed Wood* is a tribute to Ed and, on Burton's part, a labor of love. This cult favorite is also very important to the student of film. Ultimately, it is an art film. For those of us who believe in our dreams, Ed is a true hero.

To call Ed dedicated is not enough. Say what you want about his movies, the man had persistence, he had a vision, he had determination, and he never, ever, not once, thought that he couldn't "make it" in Hollywood. A lot can be learned from his resilient attitude. As we have discussed in other chapters, in the fifties, television had been introduced into the American living room and the movie industry saw a decline in attendance at the box office. Many techniques were used to get the public back in front of the large screen, including various types of color film, 3-D special effects, CinemaScope, and widescreen blockbusters. Ed thought that his outrageous B-movies would do the trick. William Castle, the director of the previously discussed *Hollywood Story,* and Roger Corman were his contemporaries and all of them dabbled in horror themes. Okay, so in his personal life Ed was a cross-dressing, angora sweater–wearing fast talker (portrayed so well by Johnny Depp), but he still believed in his work to the point of fanaticism.

Burton has taken a look at the middle of Ed's career, the time he hooks up with famous actor Bela Lugosi (Martin Landau), whom he befriends. Talk about attaching talent to your project (!). Lugosi is old, a has-been, a heroin addict, yet Ed recruits him and has him starring in his leading roles. Watch Ed as he pitches his projects ("Doctor Acula . . . Doctoracula . . . Dracula"), as he struggles to keep his marriage alive and as he tirelessly gets his films produced.

There is a quintessential scene set at Musso & Frank's Grill, a Hollywood mainstay. It takes place between Orson Welles (Vincent D'Onofrio) and Ed. They are comparing notes on filmmaking. *Both* of these men are artists in their own right—only some would argue that one is perhaps the best director of all times and the other is the worst director of all times. Yet, they are bonding on issues relating to their art. The scene relays wholeheartedly the essence of what it means to be an artist. Both of these men pursued their dreams and lived their lives as they wished, no matter what was said of their art. This is what it means to be an artist, a true director. Believe in your work and see it through. "Visions are worth fighting for," is the advice Welles puts on the table. They sure are.

A SPRAWLING SUBURB, SUPERB FOR FILMMAKING

The sixties and seventies find L.A. to be a tableau upon which young filmmakers shoot their films. The backdrop of this sprawling suburb-like city is used to establish a very American Southwest look of definitive All-American cinema. So, here, now, are the films of the sixties and seventies that feature directors and their creatively complicated industry lives.

David Holzman's Diary (Paradigm, 1967)

Writer L. M. Kit Carson stars as a young filmmaker who takes life very seriously. So seriously that he turns his camera inward to shoot his own life in an early example of cinema verite. Some viewers have watched this piece and called it self-indulgent, angst-ridden, and boring, especially when the lead is going through a romantic breakup. However, it is important to understand that this is one of the first fake documentaries and it is meant to be a mockery of student films. This film was produced on a very low budget and was considered "underground" for its time. (Think of *Blair Witch Project* and its success. This will give you an idea how important this film was in the late sixties—at least to young filmmakers—and how it was kept under the radar of any of the big studios; that fact alone gave it some cache.) The grainy black-and-white mise-en-scène only adds to how well constructed this project is—it is even shot like a student film. This is the effect that director Jim McBride wanted in this one-of-a-kind "little" film that captures the energy of the late sixties in America.

Targets (Paramount, 1968)

Ed Wood latched on to Bela Lugosi in Bela's twilight years. Director Peter Bogdanovich does the same thing in *Targets,* his directorial debut, which features a plot that echoes the outbreak of violence within society during the late sixties. In it, he portrays Sammy Michaels, a young filmmaker who convinces aged horror star Byron Orlok (Boris Karloff) to star in his thriller.

Michaels pursues and eventually persuades this once-important elderly actor to be part of his film. What Michaels doesn't know is that Orlok has a stalker who has been planning to attack him for some time. As Michaels begins production on his film, Orlok cooperates and agrees to being Michaels's star. Orlok foreshadows the plot with a twist of irony when he delivers the line "No one's afraid of a painted monster anymore . . ." after watching a convenience-store killing.

Michaels' production is underway. At the same time, Orlok's stalker, himself a victim of the violent society he lives in, begins his random killings. Bogdanovich sets out to make a statement when he presents a scene between Orlok and the sniper. The deranged sniper thinks Orlok is an old monster (in this case, a Frankenstein type, as Karloff was the original Frankenstein's monster). The sniper's intention to kill is thwarted, because this delusion frightens and confuses him.

Film buffs often find Bogdanovich's first film to be masterfully executed and admire it for revealing the effect of on-screen violence upon everyday violence. To many first-time viewers, much of the action of this film looks campy, as if it were straight out of the *Batman* television series. As an example of the director's job in Hollywood, it does show the character of Michaels and his determination and grit. He sets out to get his film made, and does just that. For that reason, *Targets* is worth a look-see. It is also a filmed document of the late-sixties era and filmmaking of that time.

Alex in Wonderland (MGM, 1970)

This very seventies piece is actually just a large snapshot of Los Angeles in 1969–1970. Paul Mazursky wrote the screenplay for this movie. Donald Sutherland stars as a young film director who has just completed his first film. The film has been heralded as a masterpiece, and now Alex must live up to his fame by presenting a second film that is at least as good if not better than his first.

Alex is experiencing his newfound success and dealing with the changes. He consults a psychiatrist, visits his friends, argues with his wife, and experiments with LSD. The highlight of this flick is when Alex pays a visit to the famous Italian film director Federico Fellini, his idol. He tries to bond with Fellini (in a scene that doesn't work as well as the bonding session between Welles and Wood in *Ed Wood*), but the elder director really doesn't want to have anything to do with the mixed-up, hippie-like Alex. Alex continues on his quest to find answers, and after an encounter with the French actress Jeanne Moreau and an assignment to direct an art film given to him by an MGM exec, he resorts to his treehouse and tries to solve his problems from there.

Alex in Wonderland is the familiar rags-to-riches scenario; it's just that this one is set in 1970 Hollywood and the fashion, the look of L.A., and the energy of the entire film are not to be missed.

Annie Hall (United Artists, 1977)

Annie Hall is included in this chapter because it is a film that depicts the feeling of show business in the seventies and accurately reflects both New York City and Los Angeles at that time. Alvy Singer (Woody Allen) is a writer-director-actor who unravels and reveals his love story with his girlfriend Annie Hall through a series of flashbacks and vignettes. Alvy says: "I thought of that old joke, this guy goes to a psychiatrist and says 'Doc, my brother's crazy. He thinks he's a chicken.' And the doctor says 'Well, why don't you turn him in?' and the guy says 'I would, but I need the eggs.' Well, I guess that pretty much how I feel about relationships. They're totally irrational and crazy and absurd and I guess we keep going through it because most of us need the eggs."

When he discusses his work, he speaks like a true American auteur, and when he is asked to visit Los Angeles, he makes his opinion of Los Angeles, which isn't good, known as he exposes the weirdness of the city at that time. He visits an industry party where starlets are plentiful and mantras are forgotten, and learns that laugh tracks are attached to most of the television comedy shows. He doesn't like the fact that one has to drive everywhere, and illustrates the weird juxtaposition of the symbols of Christmastime (a flying Santa and his reindeer hanging in the streets in Beverly Hills) against the hot sun.

Annie Hall is important to see for the difference in energy between New York City and Los Angeles . . . and for Allen's performance as Alvy, the typical director-type who is not always very sure of himself but somehow comes out on top. These opposite-energy Meccas of show business remain the same today.

THE EIGHTIES: ONE MEMORABLE, AND TWO ODD, FILMS

The following three films all take a real insider's look at this job — perhaps a little *too* "insider."

Stardust Memories (United Artists, 1980)

In Woody's next film, he matures a bit. *Stardust Memories* is essentially Woody's own personal $8\frac{1}{2}$. The film uses a film festival as the backdrop to illustrate a filmmaker's lifestyle. Its black-and-white footage makes the auteur's lifestyle seem even more celluloid-like. Filmmaker Sandy Bates (Allen) is seen dealing with his neurosis, his love life, his creative blocks, and his personal fears.

Bates attends a film festival in his honor and interacts with his public as he conducts seminars and nervously answers questions of his fans. The movie is truly an insider's look at the behind-the-scenes life of a famous film director. Here this insecure personality deals with the way the public accepts his art. Obviously, he is a very successful artist, because fame follows him; nonetheless, he is still human and has everyday fears just like every other human. This film reflects the life of a filmmaker as an artist and should be highly scrutinized by anyone wishing to live the life of an auteur. The film doesn't reveal so much the action of the job, but what to expect when one has reached a level of success as a director—a perfect film to usher in the opulent excess of the "Me" decade.

Honorable Mentions

The Legend of Lylah Claire (MGM, 1968)

Silent Movie (Twentieth Century Fox, 1976)

In *The Legend of Lylah Claire* director Lewis Zarkan (Peter Finch) takes it upon himself to mold young actress wannabe Elsa Brinkmann (Tuesday Weld) into a famous actress who had died mysteriously a few years earlier. The movie is a tweaked horror film worth watching only for the scenes of Lewis and his production team sitting around and discussing how they will transform Elsa from a fairly good-looking, bespectacled young woman into a glamorous bombshell.

Silent Movie is a spoof of silent movies that features the usual Mel Brooks madcap fun with lots of slapstick schtick and simple dialogue written in inserted frames on the screen. It is clearly the silliest film version of the job of a director.

REALITY IN THE NINETIES

The nineties bring us to total reality, especially at the end of the decade, when most of these films were made. All of them take "the man behind the curtain" attitude, exposing the truth for what it really is, and what it is like to live the director's life.

The Pickle (Columbia, 1993)

Harold Stone (Danny Aiello) is a successful feature-film director who is trying to retain his position as a working director at the top of the Hollywood Food Chain. It is the last three days before the release of his latest movie, *The Pickle,* which is about a group of farmhands who grow an enormous pickle and fly it to another planet, Cleveland. Cleveland is a strange planet where everyone only eats meat and dresses like an Upper West Sider—in black spandex and dark sunglasses. *The Pickle* is a typical Hollywood studio–produced piece of garbage, which Stone realizes. He also knows its going to bomb at the box office.

In this second film about Hollywood written by Paul Mazursky (his first was *Alex in Wonderland,* discussed earlier in this chapter), Stone is seen going through the motions of a contemporary film director. He does his pitch meetings. He deals with studio execs. He has various ex-wives and daughters to contend with. He attempts suicide during the premiere of his film, which he doesn't attend. Halfway through the attempt his phone rings. He learns that the film is a huge hit. A spaceship pickle floats away above the city and Stone lives happily ever after.

While this film is silly, ridiculous, and difficult to watch at times, Aiello in the lead portrays a very good director character, who is obviously concerned about the quality of work he releases and the audience's reaction to his work. This film also shows the human side, the reality of being a contemporary director. There are a lot of pressures, both personal and professional. Monetary problems, lack of creative control, complicated families, and suicidal tendencies are part of life in the nineties, and Harold Stone is only the beginning.

Burn, Hollywood, Burn (Hollywood Pictures, 1997)

Burn, Hollywood, Burn is sometimes referred to as *An Alan Smithee Film: Burn, Hollywood, Burn,* which immediately gives it away as the dog of a film that it is. Eric Idle stars as a director who is angry about the project he is directing—so much so that he doesn't want his name in the credits and in fact doesn't want anyone to ever know he was involved in this horrible piece of crap. Overrated screenwriter Joe Eszterhas is the creator of this film, his personal slap in the face of Hollywood, the same Hollywood that paid him millions to write mediocre screenplays such as *Sliver* and *Showgirls.*

It is a tradition in Hollywood to use the name of Alan Smithee if a director is not happy with a film, and so, Eszterhas embraced the opportunity to make a statement about creative control and produced a movie about just that type of situation. It's ironic that the real-life director of *Burn, Hollywood, Burn* ended up removing his name from the film, so it truly is an Alan Smithee film about an Alan Smithee film. It is painful to watch and only those who are die-hard director wannabes need tune in.

Gods and Monsters (Lions Gate Films, 1998)

From one of Hollywood's worse films, to one of her best . . . *Gods and Monsters* is a celebration of artistic creation and a must-see for every soul who wants to try his hand at directing. Actor Ian McKellen portrays director James Whale. Whale had a long career in Hollywood, most famous for directing *Frankenstein* and *The Bride of Frankenstein.* This story takes place in the twilight of his life.

Whale, distinguished, proper, and gay, is attracted to his gardener Clay, portrayed by a very buff Brendan Fraser. Whale pursues the young man. An interesting game of cat and mouse begins, peppered by the sarcastic wit of housekeeper Hannah (Lynn Redgrave). *Gods and Monsters* looks at Hollywood from a very realistic point of view. The absurdity of fame is revealed in a scene that shows Whale at a Sunday brunch hosted by George Cukor. Here a young publicity hound exclaims to Whale, as he gathers up Elsa Lanchester and Boris Karloff, "I could see you with your monsters!" A reunion of sorts takes place and a picture is taken—all three of the "stars" are noticeably uncomfortable due to the fact that they were famous decades ago and would prefer to keep that time in their lives in the past. "Don't you just love being famous?" they say, with irony. Whale himself knows that "the only monsters that exist are those that are in your head" as he experiences flashbacks—flashbacks from the war and from making movies—as part of his dementia. He is approaching death. This entry in the chapter about directors provides a solid dose of reality as one man grows old and lonely and realizes his fame can no longer bring him joy.

The Truman Show (Paramount, 1998)

"We've become bored with watching actors give us phony emotions," says Christo (Ed Harris), an electronic artist–type director and creator of a twenty-four-hour television program titled *The Truman Show*. "We're tired of pyrotechnics and special effects. While the world he inhabits is in some respects counterfeit, there's nothing fake about Truman himself. No scripts, no cue cards. It isn't always Shakespeare, but it's genuine. It's a life." And with this 1998 movie, the ultimate in self-examination unfolds as Truman Burbank (Jim Carrey) lives his life in a big studio, a personal lot, so to speak, not knowing that it is, in fact, a fake life. In the story, Truman learns the truth about his life and Christo as his leader.

Christo is the mastermind, the ultimate in director energy as he not only directs his lead character's life on an hourly, daily, weekly, and yearly basis, he also basically becomes the character's godlike entity, an accomplishment all directors would love to achieve. And for that reason, Christo is included here as an honor to the work of a director.

American Movie (Sony Pictures Classics, 1999)

And finally, a documentary by filmmaker Chris Smith, who attended college at the film school of the University of Wisconsin-Milwaukee and met a fellow student, named Mark Borchardt. Borchardt was toiling over a short film titled *Coven* and in preproduction of a feature-length film titled *Northwestern*. Smith didn't miss the opportunity to turn the camera on the over six-foot-tall

Borchardt and his innocent enthusiasm. The problem was that Smith entered film school and left years later and Borchardt was still editing the short film.

And Smith was right—Borchardt's life as a naïve but dedicated film-maker on Milwaukee's Northeast side provides an entertaining scenario as he makes his family work as extras, hounds his elderly uncle for financial backing, and begs his friends and lovers to help him get his film done. You can't get more real than Mark Borchardt's life seen here, in this excellent documentary, which shows what it's like to be an ordinary independent filmmaker. The film is fascinating, sometimes seemingly unreal, and yet so real that it's scary. If you want to be a director, *American Movie* is a must-see . . . if only in order to learn how *not* to get a movie made.

SEVENTEEN STORIES

These seventeen movies focus on dedicated, eccentric, and extreme direc-tors and their passion for their work. As they choose their projects and exe-cute their will on the set, they discover starlets, save studios, meet their mentors, run over budget, and even reminisce about their careers. No one can take away the fact that being a director is a very powerful position and extremely hard work—and the rewards are boundless. It's no surprise to hear that eternal mantra around Tinseltown that goes something like "Yeah, but all I really want to do is direct…"

IN REAL LIFE

Here are two brief meetings with women who are fighting for their visions.

Producer-director Lynn Woodbury on Directing

Lynn Woodbury is an independent producer-director. She is black. She is interested in feminist themes and has had success on the film-festival circuit.

What films have inspired you to be a director?

My favorite film about filmmaking is *David Holzman's Diary* by Jim McBride. I also really like *Living in Oblivion* and *The Watermelon Woman.* In all three films we go behind the scenes and experience how independent filmmaking affects the lives of the directors and how the drama in their personal lives is played out ultimately in the films they create.

What film influenced you in your work as a director?

When I became interested in filmmaking, I was living in Taiwan. It was the late eighties and the Taiwanese independent film movement was just beginning. I remember seeing a film by Edward Yang called *The Terrorizers* and being absolutely blown away by it. It loved the emotion, the nonlinear narrative, and the uncertainty of how the story would turn out. I had never seen anything like it. For me, it was art on screen.

Do you feel you are in competition with other filmmakers?

Since no one seems to be interested in telling the particular stories that I want to tell, I don't feel like I am in competition with other filmmakers at all. I actually wish more filmmakers were interested in telling stories with black women protagonists. That would be an amazing day, to see directors fighting to be the first to tell stories about black women.

Do you see a strong future for independent films?

Women filmmakers will define independent film in the future. Statistics from the Directors Guild of America suggests that the industry isn't interested in our stories or working with us. But our numbers grow stronger every day and we are more determined to tell our stories and continue to figure out more innovative ways to succeed in bringing our stories to the screen without the support of the industry.

Producer-director Kristin Meadows on Directing

Kristen Meadows is a producer-director who has just completed her first short film, a love story set in Monaco.

Have you ever been influenced by any of the movies about movies?

I've watched some of these films and yeah, I've questioned my career but I haven't switched industries yet!

What is the future of short films and indie-made projects?

I definitely see a future for both. Shorts really exploded with the expansion of the Internet, allowing people to become accustomed to their presence. I think once the technology is up to speed, shorts will have the capacity to generate almost as much attention as feature-length films. I don't think independent film will ever go away. People who want to make films will always do so, and people will want to see them. A lot of people say the movement is dead, I just think it is reinventing itself. Just look at what we have been given by the studios lately; it won't be tolerated for very much longer by audiences. And again, we will turn to the independent films.

CHAPTER 6

PRESS

CREATIVE CAREERS IN HOLLYWOOD

STATUS

DURABILITY: Keeper.

LENGTH OF STAY: Two years at a PR firm, a lifetime if it's your own company.

FOOD-CHAIN VALUE: Mid-level.

UPWARD MOBILITY: Good.

DESIRABILITY FACTOR: Somewhat high, especially among assistants and wannabe producers.

VACATION: Combine it with being on a shoot; an extra weekend; no more than two days.

SALARY: Grande Cap.

HOW EASY IT IS TO GET THIS JOB: On a scale of 1 to 10 (1 being the easiest), 5.

PREREQUISITES: Knowing how to read. The ability to capture the attention of your audience through press releases and sound bites. Befriending as many studio execs as possible. Befriending as many media reporters as possible. Being seen at all the right premieres, screenings, and A-list parties about town.

Certainly no folk hero or god has ever been known so intimately by his admirers as are the movie stars. But, of course, none of the ancient gods had publicity departments.
—Hortense Powdermaker, Hollywood: The Dream Factory

Gossip hounds. Tabloid reporters. Press agents. Publicity managers. The entire star-making machinery, from the early days of the star-filled magazine *Photoplay* to the press junkets of today, is a necessary part of Hollywood. Protective professionals doing damage control can be found at every premiere and screening in town. Why are celebrities so special and why do they need all of these handlers around them all the time?

STARS FALLING FROM THE SKY

In the early days of Hollywood, stars got their start in the theater. If an actor was not known on Broadway, then where did he come from? The Great White Way in New York was the only real major supplier of talent. Many early gossip mavens created their stars' background. Success tales had to be constructed. A star (which literally meant that the individual fell from the sky) had to have photogenic looks. She also had to be sleeping with the right person or persons and had to know all the right people in the right social circles. Perseverance, acting ability, and breaks (luck) were also necessary to achieve Hollywood fame. All the young girls who had won beauty contests in their hometowns, and thousands of others with exceptionally good looks—with and without talent—were intent on stardom. Many point to the 1919 article "Why Is a Star?" written by Frank E. Wood in *Photoplay* as the first attempt to explain the star system and the public's adoration of movie stars. The star system promoted these newly immortal individuals and made them larger than life

The period between 1910 and 1914 saw the birth of stars. The famed, or so the publicity myth averred, achieved their eminence as naturally as cream rising to the top of a bottle of unhomogenized milk. By the twenties, film performers were essentially studio-owned and -operated commodities. Studios designed star personalities, vehicles, publicity, promotion, and public appearances. Fan clubs grew. Photographers thrived. Press agents, publicity departments, and contracts were under the studio's control. The result was that each star had a "reel" life in front of the camera and a (studio-created) "real" life off screen.

Many of these stars would remain anonymous until the public would pick a favorite. Fan letters would be sent to the studio, and the popularity for each of the newly selected stars would grow. In 1909, Carl Laemmle of Universal Pictures decided to test this audience's interest and circulated the rumor that one of his actresses, Florence Lawrence, had perished in an accident. The subsequent appearances of the, in reality, very much alive actress established her as a star. Laemmle discovered firsthand that stars sold movies better than any other merchandizing tactic. And studios' publicity departments ground out biographical and other star-related material accordingly. As we will see, fan magazines and newspaper columns followed the top performers' every move. To cash in on the public's insatiable appetite for gossip, Hollywood began to look at itself, not only within these printed resources but through the early silent films whose subjects were the movies. When sound came into the picture, Hollywood continued to turn the cameras onto itself, and the demystification of Hollywood began.

As the twentieth century was gaining speed, newspapers grew and telegraph and news wire services were on the rise. Showmen, publicists, superlatives such as "the best," "the strangest," "the biggest," and "the only" became part of the hype machine. Image management and damage control fed this star-making machine. There are six films, all unique in their own ways, which illustrate the detailed and cunning work of the members of the press.

PRESS CREDENTIALS

There's no excuse today to be unaware of the upcoming movies. Numerous magazines and newspapers, Internet sites, and broadcast and cable shows promote and air not only the film's trailer but also information and publicity about the film. In addition, there are interviews with the stars, which are made available for each of these outlets. The publicity department of the studio composes all the information provided to these outlets. The specific positions are as follows:

- **Unit Publicist**—the production's liaison with the press. This person works on a movie-per-movie basis and is often on location.
- **Studio Publicist**—handles general information about all of the studio's releases and information about the studio itself.
- **Indie Publicist**—works for an independent publicity firm that handles special releases for the studio. May work with foreign or classic films that the studio is about to release. This process is known as "boutique publicity."

And the old adage remains true: any publicity is good publicity. In the following movie examples, the job of a publicist is one of walking a tight rope. In times of scandal, publicists must find a way of concealing, while at the same time revealing, the truth. And, most recently, when stars are caught drunk driving, or accused of any sexual misconduct, the publicist finds himself working in the area of damage control as he carefully doles out his information to the press, minimizing the amount of harm that could come to his client's career.

THE FILMS

Let's look at these publicity-related movies. Curiously, most of them are from the earlier part of the last century.

Hollywood Speaks (Columbia, 1932)

Gertrude Smith (Genevieve Tobin) is depressed. She has just attended a Hollywood premiere at the Chinese Theater. While standing outside of the theater, she fantasizes while placing her feet in the cement prints. In despair, she realizes that she'll never be a star. She is just about to poison herself when newspaperman Jimmy Reed (Pat O'Brien) stops her. Reed is a red-blooded, all-American guy-next-door type. Honest and trustworthy, out to get the truth. He learns of her depression and agrees to help her climb to fame.

Working in the media of the day, Reed is able to put Gertrude, now renamed Greta Swan, in touch with the right people. (He also falls in love with her along the way.) Her star rises fast. She is the talk of the town, embraced by directors and producers. Her first film is a success, but along with that success comes a scandal when the director's wife commits suicide and leaves a note stating that she killed herself because of Greta. Greta is blackmailed. At this point, the press takes control of Greta's all too short and tragic career and essentially kills it. Greta's star has fallen but she is saved when Jimmy marries her, having never stopped loving her.

Hollywood Speaks is obviously a product of the thirties. This is an example of how much power the press had. The press ruled as the audiences trusted its source and believed its stories outright. Although it is obvious to any contemporary viewer that Greta could have found ways to fight back, the screenwriters of *Hollywood Speaks* didn't give her many options. Here, the Hollywood publicity machine effectively kills her career to expunge a scandal; in later years, of course, the machine learned a much more sophisticated version of spin control.

Bombshell (MGM, 1933)

Sex goddess Jean Harlow blazes in this classic Hollywood movie. She is 100 percent movie star. Lola Burns (Harlow) is tired of all her sexy films and constant publicity. She strives to make a change in her life. She attempts to marry an individual who is later arrested by her press agent Space Hanlon (Lee Tracy). Her husband-to-be turns out to be an illegal alien. She then wants to adopt a baby, but that scheme is thwarted when her father and brother come for a visit, and the adoption agency finds her crazy household unfit for a baby. She leaves for Palm Springs and falls in love with Gifford Middleton (Franchot Tone), a snob whose family doesn't like her. She returns to the studio only to learn that Middleton's family was hired by Hanlon to force Lola to return to Hollywood and get to work. She is furious but soon discovers that she is actually in love with Hanlon.

Although the plot seems muddled, there are moments of glamour and glitz in this film. Harlow's satin sheets and luxurious makeup fill the screen, making all of us nostalgic for an earlier, more innocent time in Hollywood history. The theme of having a man rescue the star from further unfortunate publicity is repeated as it was in *Hollywood Speaks*.

That makes two films from the thirties wherein the female leads are saved from the evil press by marriage.

Hollywood Hotel (Warner Bros., 1938)

The famed director Busby Berkeley and star Dick Powell teamed together for the last of their eleven musicals. The story features saxophonist Henry Bowers (Powell), who plays with the Benny Goodman Orchestra. Bowers accompanies demanding star Mona Marshall (Lola Lane) to a Hollywood premiere after winning a talent contest and a contract with fictional All Star Pictures. Once he arrives in Tinseltown, he learns of a popular radio program titled *Hollywood Hotel*, which is hosted by columnist Louella Parsons. This movie is primarily a musical, as Bowers is thrown into madcap musical mischief. The show is just one big promotional piece for the town of Hollywood.

The famous song *Hooray for Hollywood* is played to death. The show within the movies is promoted as "The Rodeo of Radio, the Mardi Gras of Movieland." "It will turn Hollywood thrill side out, funny side up." Filmland will never be the same as Hollywood toots its own horn (quite literally) in this thirties musical classic. All these years later, it seems a little corny, but you have to hand it to them, they knew how to promote themselves. This end-of-an-era musical surely provides the last hurrah, as the entire film is merely a vehicle to promote the Benny Goodman Orchestra, the hotel the film takes place in, and Hollywood itself. The lack of storyline and character development assures that the promotional aspect of the film is in the forefront. Busby just wanted to make sure the rest of the world knew Hollywood was alive and flourishing.

Affairs of Annabel (RKO, 1938)

Much can be learned from Jack Oakie's performance as an overactive, super-hyper press agent in this delightful movie from the thirties. Lanny Morgan (Oakie) is a press agent for Wonder Pictures. He's got a tough job in front of him. Annabel Allison (played by Lucille Ball in what is surely an early prototype of her signature character in the *I Love Lucy* show), is an out-of-work actress who needs a movie hit. In the past, in order to promote Annabel, Lanny has thought of some outrageous things, such as having her spend time in jail for an upcoming prison movie. The next project involves the lead character of a maid, so Lanny arranges for Annabel to be a domestic caregiver and hilarity ensues as Annabel learns her new job. When her next assignment appears and it involves smugglers, Annabel finds herself in jail again, and so the wacky wheel of publicity churns on and on.

Lanny is the ultimate PR professional. He thinks of ways to promote his client. He's good; in fact, he's the best of the fast-talking "you know me, always kidding!" type of press agent who can talk himself in and out of everything. And on top of that, both of the leads are a joy to watch in this nearly perfect sketch-comedy movie.

Beloved Infidel (Twentieth Century Fox, 1959)

Beloved Infidel features actress Deborah Kerr as British-born gossip columnist Sheilah Graham. This movie is a glossed-over version of her autobiographical book. She is seen arriving in New York from her homeland, Britain, and then eventually moving to and working in Hollywood. She is the perfect paradigm of the female writer of the time. Like her contemporaries Hedda Hopper and Louella Parsons, Sheilah has a confidence about her. She is a brazen image of strong femininity. One of her first assignments involves walking into a studio soundstage where a production is in progress. Here she comes face to face with an actress she has panned in an earlier article, referring to the actress as a "witch with a capital B." In the confrontation that follows, Sheilah doesn't back down.

Kerr portrays Sheilah brilliantly. When she and F. Scott Fitzgerald (Gregory Peck) meet at a dinner party of a mutual friend, she is coy, coquettish even, yet radiating a strong aura of intelligence. No wonder Fitzgerald found her attractive. Their love affair grows as he teaches her about famous fiction writers and writing and she jet-sets around town to screenings and events. Hers is a glamorous life that is shattered when Fitzgerald's alcohol addiction worsens, and he starts to appear drunk in public places. Being no stranger to masking the truth about others, she is faced with having to cover up the truth about her own life.

Hollywood publicists and press reporters (nowadays joined by television reporters) continue to live the same life seen here. They are frequently deep in the middle of every promotion, running from premiere to premiere to organize interviews with their clients or representing films as they open. This is a hectic life and one that requires being a watchdog of the media. The competition is tough, so your eyes need to be not only on your own client but also on all the other stars—and on how their handlers promote their clients or out-promote yours. There is a hint of competition in this film, a hint of "covering up" the truth; however, with the next film, the method of spin and damage control sets the pace and establishes the precedent for the future of the public-relations industry.

The Big Knife (United Artists, 1955)

Based on Clifford Odets's 1949 Broadway play of the same name, this black-and-white film plays like something from Masterpiece Theater or an extended *Twilight Zone* episode. The action of the story takes place in one location—the home of a major movie star Charlie Castle (Jack Palance, who overpowers this story by his extremely tall frame alone). It seems that Charlie is upset with the Hollywood system that is being forced down his throat. His entire life is played out according to the dictates of the top brass of his studio—Stanley Hoff (Rod Steiger) and his hatchet man Smiley Coy (great name, played by Wendell Corey). Charlie has become a bored, superficial player of meaningless parts but Hoff won't let him out of his contract. When Charlie says he won't sign his contract, Hoff reminds him of the publicity machine that could destroy him, and sends arrogant publicist Patty Benedict (Ilka Chase) to interview him. Charlie is aware of the damage she could do to his already tired relationship with his wife (Ida Lupino) and his career, but he still refuses to sign, and this is when Hoff and Company storm in.

It seems Hoff knows of a past incident involving Charlie. He was driving drunk and killed a young girl. Hoff had another studio lackey serve the time, leaving his prized star free to make money for his studio. Charlie knows that if he doesn't sign, Patty Benedict will be the first to know the truth. His hands are clearly tied. The truth, which will spread like wildfire through the publicity desks of the studio and all of the Hollywood outlets, would kill him. He takes matters in his own hands and kills himself. Hoff and Smiley Coy see to it that the truth is never revealed and use that same publicity machine to fabricate the story of Charlie's death. Charlie's a commodity, protected by the studio after all.

Odets set out to write a story about a popular star at odds with the Hollywood system and set up a studio head to be his main villain. It is an interesting fact that this film was released through United Artists, a production entity known as a refuge for independent filmmakers of the time.

Though not a box-office success, *The Big Knife* has remained an audience favorite. It defines the control and power a studio had over the star. That this story showed the studio covering up a scandal was edgy for 1955.

Today, spin control is the most used form of publicity. The truth, if it might offend the public and therefore hurt box-office dollars, will be hidden, masked, protected at all cost. This is a tricky job, one that demands a shapeshifter mind and a brilliant sense of spin. It is also one that requires the individual to be loyal—and yet not always faithful to the truth. The publicist must be faithful to the constructed truth and stand by it, wholeheartedly.

BEFORE E! ENTERTAINMENT TELEVISION

Of the six films that feature characters working in the publicity machine, four take place in the thirties and two in the fifties. Curiously, this was a profession that held much more of a cachet in the early days of Hollywood than it did during the last half of the century and in current times. The Louella Parsons, Hedda Hoppers, and Sheilah Grahams of the world tended to lose their edge after the fifties. Their type of reporting became ordinary and average with the arrival of television. With the onslaught of cable, entire channels are now dedicated to entertainment alone. Twenty-four-hour programming about stars and gossip and pop-cultural items are now very accessible.

Celebrities are and always have been nothing but images. Today celebrity is everywhere and is as instant as coffee and available to anyone—good or bad—who wants to make some sort of statement in society. The cult of celebrity has become an obsession with the public, and the men and women who work the press do a mighty fine job of controlling their clients and protecting their assets.

PUBLICITY, PUBLICITY, PUBLICITY

Of the films explored in this chapter, perhaps *Bombshell* and *Affairs of Annabel* are precursors to the mega-publicity machine created in the late twentieth century, as lead characters go all out with outrageous schemes to promote their celebrity and their films.

However, of all of these films, *Beloved Infidel,* in its quietness, shows the life of a publicist and her triumphs, trials, and tribulations, and best focuses on the job of a publicist. What remains true, however, and is reflected in all of these films, is that the publicist and publicity department are fierce. The publicity machine stands by its product, often utilizing every spin factor available to best protect the reputation and honor of their client. As anthropologist Hortense Powdermaker stated in the opening quote of this chapter, no folk hero or god has ever been so intimately known by his admirers as are movie stars, and that is solely because movie stars have publicity departments.

Studios run the publicity machine in the twenty-first century with gala premieres, domestic and international press junkets, and promotional gimmicks for both the industry insiders and the public (i.e., t-shirts, hats, keychains, and merchandizing of all sorts—including the bag of popcorn you buy at the theater). It is often said that the more trinkets, printed material, and tchotchkes produced per movie, the worse the film is—and that if the premiere is gaudy and over-the-top, then the film also lacks in substance. Some studios and agents beef up the product—make the outside of the house look great so you'll come to the door to see inside (or see the movie)—but once you do, you realize the "inside" doesn't look very good. Smoke and mirrors make up a great deal of today's publicity machine, not all that differently from the early days and the examples depicted in the films we've looked at here—it's just done on a larger, grander scale.

IN REAL LIFE

The job of publicist will always exist. As long as there are movies and stars to promote, there will be publicists. Here are two contemporary viewpoints on this very important creative career in Hollywood.

Two Publicists, Two Points of View

One's a publicist for a small boutique PR agency. His name is Michael. The other works independently, and has worked on some of the best-known films of the past decade. Her name is Louise. Let's see what each has to say about working as press agents. Michael is interviewed first.

Michael

On influences:

When I was young, TV and film stars were my idols and I wanted to be able to work amongst them and this was my opportunity. Some of my favorite films were *The Sound of Music*, *16 Candles*, *Grease*, and, of course, many of the classics.

On what works and what doesn't work:

After working in entertainment for over ten years, I have been able to notice what works and doesn't work. Usually, projects work if there are 1) hot stars (e.g., Julia Roberts); 2) hot stars who appeal to teens (e.g., Freddie Prinze Jr.); and 3) well-reviewed films that sometimes take a little while to build an audience (*The Sixth Sense, The Others*).

I worked on *Circle of Friends* some years ago. It starred Minnie Driver in her first feature role in a film released in the United States. At the press day, it was obvious that she had the spark that it takes to be a star. In the office, we commented to each other that she seemed to have the charisma needed to make it in the business, and we were right. What followed soon after was *Good Will Hunting*, for which she was nominated for an Oscar. So, for me that was an example of being able to see what people talk about when they say someone has "It."

On competition:

There is a lot of competition in public relations. We work with the same clients on a regular basis, but there are many other agencies in town, which are connected with the other distributors. We do work on the occasional film from other distributors, but we have certain loyal clients that keep us very busy. Additionally, we work with the Hollywood Foreign Press often as studio liaisons on their award campaigns, and as we have a good relationship with the Hollywood Foreign Press Association, it has become our forte.

On working at a boutique agency:

The advantages of working at a small boutique agency are that we are able to make choices. We can choose whether or not to take a film; we don't have to take everything in order to pay the staff, rent, etc. That is the main difference. Additionally, my boss has implemented a civilized work schedule, in that we don't work too many late nights or weekends, and this is unheard of at any other agency or studio PR department. Being smaller also lets us have a more hands-on approach with each film and filmmaker. They are not apt to get lost, like within the larger companies.

Louise

On influences:

So much of Hollywood influenced my life; I sometimes wonder what is real and what I've tried to imitate. Is it fantasy or reality—for me it's blurred, because my entire life has been a reaction to Hollywood. If I've been inspired or excited, it's because some film moved me to another plane. The films that influenced me were *To Catch a Thief, Brief Encounter, Singin' in the Rain, Cinema Paradiso,* and *The Sound of Music,* to name just a few in a very long list of favorites. Hollywood has been so influential to me that I have fallen in love with who I thought the people were, collected memorabilia about them, bonded with dead icons, and even copied or tried out things they did when they were younger because I thought it was so awesome. The movies saved me from drugs and sex when I was younger, because I followed the scripts of my leading ladies. For instance, I remember dressing up in pajamas and going to see a movie with just an over-coat on, because I read somewhere that Grace Kelly did that with her friends when she was young. To me, that was much cooler than getting high.

On what works and what doesn't work:

I've never seen myself as someone who can predict what audiences will like in the future because I like to reinvent the past in my press campaigns, and Hollywood generally hates the past. This is more of a nostalgic approach, which makes me an expert in these types of films but not in blockbusters. I can some-times predict where an aspect of our culture will move, but in terms of betting on the next big thing and so on, no. But I can predict when a movie comes out if it will be a success. First, because I can watch how something is marketed, manipulated, and promoted, and then I can tell how it reaches me and my friends, family, and so on.

On competition:

I am not a competitive player. I do my own thing even if it is unpopular. I believe in what I do, because when I believe in what I am doing it affects me deeply and it becomes the sole purpose in my life. I've met with many old Hollywood players and that's where I've really learned about the biz. That is also where I've created my niche, my expertise. I fit into that market and begin to work from there. Also, I have tenacity and determination. I will never give up, never get depressed by the rejections I get. I find success in everything I do and applaud myself for how much I've tried.

On working alone:

Working on my own and for myself benefits me the most. I consider myself a maverick; I knew I didn't fit into any category or kind of company; and I wanted to teach myself the industry differently, using the method I created. For a long time I submerged myself in as many books, biographies, and films as I could watch about TV and film, in writing, and in meeting older professionals I admired and asking them questions. I like the term "trailblazer"—after all, this industry is all about being on a particular journey or trail and setting your own course. Working for myself is best for me, because I benefit most—all the efforts I put in my projects pay off, because I see everything through and can oversee all aspects.

CHAPTER 7

PRODUCER

CREATIVE CAREERS IN HOLLYWOOD

STATUS

DURABILITY: Either two to three years of indie-madness or a lifetime.

LENGTH OF STAY: See Durability.

FOOD-CHAIN VALUE: Upper echelon.

UPWARD MOBILITY: You're already there.

DESIRABILITY FACTOR: High.

VACATION: Sure (if you are producing your fifth $50-million-plus opening-weekend hit).

SALARY: Venti black coffee, straight up.

HOW EASY IT IS TO GET THIS JOB: On a scale of 1 to 10 (1 being the easiest), 1. Anyone can be a producer—you just have to have stamina.

PREREQUISITES: Anticipating what the public wants a year ahead of time, being well connected, and knowing when to dial up the right talent to pull off the next big-opening-weekend blockbuster.

CHILI: I'm going into the movie business. I'm thinking about producing.

TOMMY: What the *&#% do you know about making movies?

CHILI: Well, I don't think the producer has to know much.

—A few lines of dialogue from Get Shorty!

The term "producer" can mean many different things in Hollywood, where there are as many producers as there are vice presidents. Everywhere you go—restaurants, gyms, nightclubs, coffee shops, swap meets, or golf courses, everyone is either writing a script or producing a project. From large studio-issued productions that need a myriad of producers per film, to independent producers who work solo and usually on shoe-string budgets, being a producer is a highly desirable position that allows, for the most part, creative control over the project. On the Hollywood Food Chain, this position ranks high. Being a successful producer can be one of the most rewarding jobs in this industry.

THEY ARE ALL VISIONARIES

Many producers, especially those who have had the opportunity to create and build a body of work, become known for the type of film they produce. Modern-day producers are mini-versions of the great movie moguls. There are eight films that explore the role of producer. Each film spotlights the job of producer and the role he or she may play in getting a movie made. In the best-case scenarios, producers are often referred to as visionaries; in other circles, they can be called B-movie makers.

JUST WHAT DOES A PRODUCER DO?

The term "producer" can mean a myriad of things. Watch the credits of any given film; you'll see listings that include executive producer, associate producer, line producer, and the age-old generic producer. What they mean depends on each film. These are mutable titles. What an associate producer does on one film may be what an executive producer does on another films. With an independent film, there may be only one producer who dutifully performs all of the tasks that would require several producers on larger films.

The Producer Breakdown

- **Executive Producer**—this is the man with the money. Typically, an executive producer will either plop down the money for a project or find the money to do the project. He is usually not involved in the creative aspects of the piece. This could also be a studio head—the one of Miramax, New Line, Lion's Gate, et al. will often take that credit—and often the exec who oversaw the production will get that as well. Or it could be a manager who got a big star to say "yes."

- **Creative Producer**—the auteur of the project, the person who discovers an idea, packages the project with all of the appropriate elements (writer, director, and so on), and then goes on to sell the idea to a studio, or production company, or any number of investors to make the film independently.

- **Indie Producer**—an independent producer is one who does not have any affiliations with a studio or production company and basically works on an independent basis, gathering the money and elements on his own to get the project made. Often, after the indie producer has one or two moderate to well-received movies under his belt, he is offered a deal with a studio or production company and becomes a producer attached to an entity, indie no more. A creative producer and an independent producer can be or do the same thing. some are attached to studios with producing deals, and some aren't

- **Line Producer**—this is the hands-on producer. He is the person who organizes the practical aspects of a production to actually make it happen.

- **Associate Producer**—these are glorified titles for individuals who assist the executive, creative, and line producers, or who assist the director or star. Many of the individuals in these positions are literally assistants who have moved up a notch on the Hollywood Food Chain and are available to assist the other producers with any tasks they might need help with. This is a very nebulous title that could mean just about anything as far as the tasks that are actually performed.

- **Producer**—any individual who works at putting a project together at any level in its development. This person produces. Plain and simple, he produces.

ME? A PRODUCER?

Just about anyone can be a producer. As long as you have a nose for what the audience wants, or you decide you want to create a movie that will capture the audience's attention, you'll succeed in this very creative job. You must be willing, however, to spend a great deal of time getting started and be willing to work alone, at least to start. Producers are lone wolves. They are the only Hollywood group of the major film credit-listings who do not have a guild to unify and guide them, like the Writers Guild or Actors Guild. Generally, producers want to be alone in their pursuit of the shaping of popular culture.

In order to be a producer you just need to call yourself a producer. You are a producer once you begin to look for and gather projects that you wish to finance and develop. Get started by finding a project that you feel will appeal to the current marketplace and ultimately make you a great deal of money. Some people like to play the "passionate" card at this point in the game. "It's a project that I'm passionate about" is an oft-heard phrase. Once you discover your first story (hopefully, one that you are passionate about) that will bring you success, you can go on to produce project after project, because you will have proven that you know what the audiences want.

Producing is a job that requires knowing how to negotiate. Everyone must come away from the table satisfied—even if some are only slightly satisfied. The successful producer will enjoy power, a very nice thing to have while working in Hollywood.

Here's a look at some of the films that have featured producers in the key roles. We'll begin in the thirties and end in the nineties, exposing the extraordinary length producers will go to to assure that their passion winds up on the big screen.

Once in a Lifetime (Universal, 1932)

Based on a play that severely blasts the motion-picture industry and its way of doing business, *Once in a Lifetime* features three leads who are traveling westward on a train to try their luck in films. These three, George (Jack Oakie), May (Aline MacMahon), and Jerry (Russell Hopton), are former vaudeville performers who decide to present themselves as experts on voice culture, something they do not know anything about, in order to find new jobs once they arrive in Hollywood. (This plot could also be known as "Audio Killed the Vaudeville Stars"—keep in mind the shift between silent films and the talkies is taking place.) While on the train, they meet a Hollywood gossip columnist, and through her, they connive to find the necessary industry connections.

Once in Los Angeles, they meet with a studio mogul Herman Glogauer (played by Gregory Ratoff, who played similar roles in *The Great Profile* and *All About Eve*). They meet the usual Hollywood studio-lot types—a writer, some loony actresses, and a deadpan receptionist who repeatedly states: "Mr. Glogauer is in conference" when asked absolutely any question at all. George, May, and Jerry find themselves in Glogauer's office, where George repeats words a writer has just spoken to him: "Hollywood is run by a batch of fools." Glogauer loves the statement, much to the trio's surprise, and makes George a supervising producer on the studio lot.

George, really just a dimwitted braggart, makes a number of decisions on the lot based on no experience whatsoever, and is blithely unaware of the mistakes he may or may not be making. In situation after situation, George makes a haphazard decision and somehow comes out on top. He orders two hundred planes for a scene, which at first seems a financial disaster, until the studio realizes it can make money renting them out to other studios when they're not being used on the Glogauer lot. Every absurd mistake this man makes is interpreted as inspiration or innovation.

While *Once in a Lifetime* does present a satiric portrait of Hollywood, its characters are archetypal. Jack Oakie's portrayal of this silly character is truly a prototype for Jerry Lewis's Tashman character in *The Errand Boy*, a film that would appear over twenty years later. George proves that anyone, simply anyone, can become a producer. No experience is necessary. Soberer teachers may stress that a good producer needs to have his fingers on the public's pulse, but the lesson of this film is clear: one does not need previous experience or even an education to be a producer.

Stand-in (United Artists, 1937)

The sale of the Colossal Film Company is the topic of this film, which focuses on the corporate side of Hollywood filmmaking. Leslie Howard stars as Atterbury Dodd, a financial wizard called to the West Coast from a New York bank to track Colossal's financial performance. Dodd is escorted around the company by Miss Lester Plum (Joan Blondell), who is a stand-in for the company's star Thelma Cheri (Maria Shelton). Miss Cheri is the love interest of washed-up, alcoholic producer Douglas Quintain (Humphrey Bogart). The company is about to go under, but Dodd persuades all of the workers to seize the lot for forty-eight hours. Quintain sobers up and reedits his last film starring Thelma. The film is a success, and everyone winds up "happily ever after."

This film is heavy with dialogue that echoes corporatespeak. In its overall execution, it is monotonous; however, it does provide a number of black-and-white montages of the way things used to be in Hollywood. The Derby, Crossroads of the World, the world's first outside mall, Tail o' the Pup hot-dog stand, the corner of Hollywood and Highland complete one montage. The second one reveals a 1937 Hollywood and her top entertainment lounges—Café Trocadero, The Victor Hugo, Bali, the Biltmore, new Florentine Room, and Café Lamaze. A splendid view of the

Hollywood where the producer Quintain had to live. No wonder he enjoyed drinking!

Bogart never really gets to shine in this film, as it is one of his earlier credits. In Hollywood, money is always a concern, as it is in *Stand-in,* but when new regimes take over, or the upper echelon puts an end to the gravy train, the workers will prevail. In forty-eight hours, Quintain finishes his movie. He proves himself to be a producer who is not afraid to be hands-on. Sometimes, this is in fact part of this job. The producer must be sure that the film is getting made—even if he is the one who has to do all of the work.

The Producers (MGM/UA, 1968)

This 1968 original movie, the precursor of the currently popular Broadway stage play about the planning of a Broadway stage play, is included in this chapter because it very nicely describes the job of a producer. Creator Mel Brooks earned an Academy Award for Best Story and Screenplay. This is the classic tale of conniving producer Max Bialystock (Zero Mostel) and his accountant Leo Bloom (Gene Wilder), who set out to put on a production so dreadful and outrageous that it will close immediately, leaving them with all of the money they raised.

"Under the right circumstances, a producer could make more money with a flop than with a hit," Leo explains to Max. Max's usual method of raising money has been to woo rich old widows. Max talks Leo into going in on the scheme with him and the two begin their search for a surefire flop. Surrounded by scripts, they search for the worst play ever written until they find the fabulous flop *Springtime for Hitler.* The team track down the writer, attach a director, conduct their casting calls, and Max even manages to buy himself a "toy" in the form of a beautiful Swedish secretary who doesn't speak English.

Things are going very well, the reviewers have been bribed, and the worst play ever written is going up for its premiere showing. Max and Leo know they've got a bomb, and venture into the nearby bar to toast their failure . . . until the audience leaves the theater en masse praising the production, screaming for more. *Springtime for Hitler* is a surefire *hit;* therefore, Max and

Leo have to pay off their patrons, a task they are unable to perform. Hence, they find themselves behind bars, but that doesn't stop them as they plan their next big production for their cellmates, their new captive audience.

The Producers is near perfect in plot and execution. If you want to know what a producer does, this is the quintessential film to see.

S.O.B. (Lorimar, 1981)

Based on truth, but mostly satire, *S.O.B.* is director Blake Edwards's shame-to-fame film. Produced in 1981, *S.O.B.* tells the story of Felix Farmer (character actor Richard Mulligan), who tries to salvage his movie *Night Wind* after it earns the worst box-office opening in Capital Pictures' history. Farmer has a multitude of problems—the president of the studio wants to recut the movie, his wife (Julie Andrews) wants to divorce him, and he has failed at committing suicide. Farmer's way out of this mess is to get his wife, star of G-rated filmdom, to perform in a highly erotic adaptation of the movie. Basically, he wants to get her to "show her boobies" (and that's a direct quote from the film). His wife concedes after much persuasion and medication, and the film is reshot. The president of the studio, however, finds a loophole in the contracts and gets the rights back for the picture. Felix tries to chase the film down, quite literally, in an exaggerated car chase that ends in a fatal standoff, leaving Felix sprawling dead on his reels.

Yes, the bulk of this film is satire and features a weighty cast—Bill Holden (in his last film before his death), Robert Preston, Shelley Winters, Robert Vaughn, Larry Hagman, Stuart Margolin, Loretta Swit, Craig Stevens, Larry Storch, Robert Loggia, Rosanna Arquette, and Marisa Berenson, but it is based on truth for the most part. Here is a producer who incessantly rants and raves. He is surrounded by the prototypical accoutrements of Hollywood producers. He has the typical divorce lawyer, the studio chief, the irate soon-to-be-ex wife, and agents and press people all around. A dud at the box office, it has been said that *S.O.B.* is Edwards's statement about the movie business after he was bashed around in the press during the seventies.

Edwards portrays the upper crust of the industry as cheating, lying, deceptive scoundrels, and the performances are over-the-top. The issues that concern the characters in this film—legal and financial rights to properties and profits—are of great concern to the real-life players in contemporary Hollywood. And the resorting to selling out to sex to make a film saleable is a very real situation.

S.O.B.'s depiction of a producer's life is pretty accurate. It's a flurry of phone calls and quick deals. Producers must trust their instincts when packaging a project, and they should believe in their own bullshit sometimes. Farmer shouts: "Even if I'm wrong, and I'm not, I'm full of fire, I'm a blazing comet." His partner is quick to comment: "Comets burn out, my friend."

. . . **And God Spoke** (Live Entertainment, 1993)

The Big Picture meets *Spinal Tap* equals . . . *And God Spoke.* In this mockumentary, director Clive Walton and producer Marvin Handleman candidly introduce this film and appear throughout, talking about how they met in film school and about their not-so-successful attempts at making movies. Their first idea, *Dial S for Sex,* didn't really go anywhere (but it was a hit in Bangladesh), hence their new idea about doing a low-budget film about the Bible. Marvin is excited, because from a producer's point of view, you cannot lose with the subject matter. He figures about four billion people have read the book, so . . . that amount times an average $7 ticket . . . think about it. And even though this film is a spoof, it does offer some very real scenarios about the adventures producers have when pulling together movie projects.

The first hurdle is getting studio backing. Clive and Marvin manage to secure a ten-million-dollar budget with a young studio executive. Preproduction begins, and cast and crew are auditioned. Tattooed Eves are disqualified, some arks won't fit onto the soundstage, and other technical snafus take place. The top brass then decides to pull the funding for the project. Clive and Marvin are hell-bent on going forward, and turn to their own families and friends, and corporations such as Coca-Cola, to raise the money. Clive quits over the product placement of Coke cans in the Ten Commandments scene, obviously a very funny statement about how far product placement has gone in mainstream moviemaking. Marvin, however, scrapes up enough money to wrap the film, and the time has come for the public to decide if God should have spoken or kept quiet.

The makers end the film on a happy note, stating that . . . *And God Spoke* bombed at the box office but resurfaced a few years later to become a cult hit not unlike *Rocky Horror Picture Show* and wound up making $42 million.

. . . *And God Spoke* does poke fun at the art of moviemaking and features a few too many stereotypes in its storyline; however, as far as producer Marvin goes, he's a pretty good example. His ego is strong enough, his determination is in place, and he doesn't lose his sense of humor during the production. This delightful little low-budget gem succeeds on many levels and is one of the least depressing views of Hollywood from within many of these movies about movies. Wanna be a producer? Marvin is a good role model.

Get Shorty! (MGM, 1995)

The film is based on Elmore Leonard's novel of the same name. Chili Palmer (John Travolta) is a Miami loan shark and an avid movie fan who is sent to Los Angeles to collect a debt from Leo, a dry cleaner who scammed $300,000 from an airline company. In order to begin this assignment, Chili must first go to Hollywood to collect another debt, from Harry Zimm (Gene Hackman). There are many creative careers portrayed in this cornucopia of Hollywood excess, but the role of producer may be seen through Zimm's outlandish behavior and Chili's attempts to become a producer. Two opportunities in one.

Zimm is a B-movie producer specializing in horror flicks. He's past middle age and is still waiting for the script that will make his career. He's a sleazy guy who is focused solely on money; after all, film is a business and it exists to make money, like any other business. To Zimm, the idea of making a film has nothing to do with art or quality, and its success is defined by its box-office gross. Self-absorbed Zimm knows how to schmooze. He personifies the stereotype of a fat-cat producer, driving around in a Rolls Royce, wearing sunglasses and a gold necklace. He's ready to say anything to make a deal, a kind of bumbling idiot who is scared of his angry investors and wants to make a blockbuster that'll bring him some money.

As the plot thickens, Zimm manages to get involved in a murder investigation while Chili is grooving on Hollywood. His "tell-it-like-it-is" and "say-only-what-you-have-to" attitude takes Hollywood by storm. He knows nothing about the movie business, except that you don't need to know much to be a producer. He just puts on his image and convinces people to jump on board his production. He walks around dressed in black and eternally wearing sunglasses, saying only what he has to and being very straightforward about it. He is quick to judge who's an idiot and who is competent, and he is always correct. Travolta's Chili is a badass, intimidating and self-confident. Before he knows it, he's making deals with major stars and he successfully manages to rid himself of the evil investors who are after him. Chili produces a blockbuster film—and he also gets the girl.

Get Shorty! is only a bit of an exaggeration. Here, "who you know, not what you know" is the number-one rule to being successful. In addition, Chili has been learning his own truths about being a producer by just hanging around. He is very likable, as is Harry Zimm. Both know how to pitch well, cut deals, and be ruthless. They are excellent representations of the sleazy side of Hollywood producing.

Bowfinger (Universal, 1999)

Last, but not least, Steve Martin's *Bowfinger*, released in August of 1999. Steve Martin has great affection for this subject, and that affection pours out on the screen as he stars as run-down actor-producer-director Robert K. "Bobby" Bowfinger. He decides he must take one last shot at fame and fortune and attach himself to a script that a friend of his has written. A successful executive-producer friend of his tells him he'll back the film if major star Kit Ramsey (Eddie Murphy) is attached. Bowfinger makes an attempt to get Kit, who says no. Bobby is then forced to shoot the film indie-style, and that's when the adventurous fun begins.

Bowfinger executes the job of producing in a perfect independent manner. He gathers up the cheapest crew possible—illegal aliens from Mexico; a fresh-off-the-boat young beauty, Daisy (Heather Graham); a diva, Carol (Christine Baranski), as his head of development and production; and a gofer production assistant and a look-alike Kit Ramsey named Jiff Ramsey (also played by Eddie Murphy) in his lead. Jiff doesn't have a clue about how to act, but Bowfinger keeps him in line. Bowfinger does his best to keep the production above water—and succeeds. Daisy has one of the best lines, summarizing the feelings of most newly-arrived-in-L.A. industry hopefuls, when she says: "I know what's going on. I may be from Ohio. But I'm not *from* Ohio." (Emphasis on the fact that she may be physically from that part of the world but mentally, emotionally, she's light years ahead of her former classmates and neighborhood friends.)

Overall, *Bowfinger* is a delightful film, chock-full of illustrations of industry jobs; however, Steve Martin's performance as Bobby Bowfinger is one that is not to be missed, especially if you are choosing a career as a producer.

Honorable Mention

Matinee (Universal, 1993)

Lawrence Woolsey (John Goodman) is a larger-than-life B-movie producer famous for schlock techniques such as Atomovision. This Joe Dante–directed film takes place in 1962 in Key West during the Cuban Missile Crisis and is really the story of the teenagers and their families who are serving in the military during this trying time. The film does offer an interesting glimpse at the world of producing, and captures what a producer's role was like during that era when it was necessary to drive across the country and do promotion in each and every market. The Woolsey character is based on director William Castle (who is mentioned in chapter 5). Ultimately, the movie is a nostalgic study of this time in American history. Goodman manages to make his sleazy shapeshifter character actually quite likable as he pursues his audience, trying to get them to escape their anxiety about the national crisis and watch his movie.

PRODUCER WRAP

Of the eight films focusing on the job of producing, at least three, *Once in a Lifetime*, *The Producers*, and *Get Shorty!* portray individuals who are thrown into producing and run with the baton to see if they can survive the race.

The other five movies, *Stand-in, Matinee, S.O.B., . . . And God Spoke,* and *Bowfinger* are about people who have designated themselves producers and who throw themselves into the race under any circumstances and against all odds. These examples illustrate that the act of producing is the easiest and the hardest of all of the creative careers in Hollywood.

Easiest because, as mentioned above, anyone can do it, you just have to say you are going to do it (of course it helps if you have some money, some backing, some "other people's money" support). Hardest because until you have your first hit (modest or grand) you remain a producer in your own head only. Producers need to have hits; they need to prove their worth in the marketplace in order to have a lucrative and successful career.

What is the key to being a successful producer? What are some lessons to be learned from the experience of these on-screen producers? Find projects you are passionate about. Ask yourself what types of projects would entice you to get out of your house and pay money to see. Hopefully, your taste, your passion will resonate to your generation and the collective consciousness. Your projects will find an audience.

IN REAL LIFE

Successful and now retired from producing, here are two real-life accounts of the same job.

Interview with Steve Cars, Producer

Steve Cars is a working producer. He's agreed to share a slice of his life with us and give us some insight into just what a producer does. We began the interview with our usual question regarding whether or not there were any movies that influenced him and his decision to be a producer when he was growing up. Here's what he said.

On influences:

I've just always loved movies and love few things better than seeing a good one. Movies really move me in a way that theater, for example, very often doesn't. The first movie I can remember seeing in a theater was either *Mary Poppins* or *Santa*

Claus Conquers the Martians (both 1964). After that, I moved on to more sophisticated fare. I loved horror films when I was growing up—both Hammer and Universal. I guess some of the movies that really made me love film (it's a really long list) are, in no particular order, *Psycho*, *The Conversation*, *The Godfather I and II*, *Klute*, *Whatever Happened to Baby Jane?*, *Chinatown*, *Seven Beauties*, *Rosemary's Baby*, *Carrie*, *Women in Love*, *American Gigolo*, *The Blue Angel*, *Days of Wine and Roses*, *The Birds*, *Citizen Kane*, *High and Low*, *La Dolce Vita*, *The Wizard of Oz*, *The Graduate*, *Valley of the Dolls*, *Who's Afraid of Virginia Woolf?*... there are so many more. So many I've seen as an adult that I would have been proud to be involved in. That's why I want to be a producer, I can't imagine anything more fulfilling than producing a really great movie that moves and entertains people, makes them think—examine their lives and the lives of others.

On future success:

I suppose there are people who can automatically know if a project or a person is going to be a success, but I think they're rare—and lucky. Having been an acquisitions executive, I guess I had to predict what the public wanted to see. Sometimes I was right—*The Crying Game* and *The Crow* (two movies no one else wanted) became hits while some that I told the company not to buy (forgettable titles like *The Opposite Sex* and *The Fortress*, et al) weren't successful at all. Conversely, I got severely trounced for not recommending *Howard's End*. I'm probably better with actors.

On the competition factor:

Producing is *highly* competitive. So is being an executive. This business is largely based on who your friends are. And a lot of friendships are based on what you can do for each other. I know executives who have been told by their bosses to make sure they go on skiing trips, spend holidays, weekends, et cetera with agents—become their best friends to get the best material or access to the best talent. For a producer, it's about making sure you're close to 1) agents who believe that if they give you a script that excites you, you'll be able to either set it up at a studio or find independent financing for it—and who will give you all their best material instead of giving it to the guy down the street; 2) executives who like you, your taste, and think you can make successful movies—be involved start to finish, i.e., finding the script, book, etc., developing it, choosing director and cast, and overseeing the making of it; 3) writers who like and trust you and know that you won't fuck them—and to whom you can go if you find a script that needs work, or a book to adapt, and hopefully get them to work for less than their quote; 4) directors who, like writers, want to work with you because they know you'll protect them (from meddlesome studio executives, nervous financiers, etc.) and allow them to make their movie; and 5) other producers who you want to partner with and who want to partner with you if you bring them a great piece of material, because you feel they can help you get it made (through their studio or financing deal or because they have a relationship with a director or star who would get it made). The film business is a jigsaw puzzle in which all the pieces better fit together.

On mentors and inspiration:

The old movie moguls (Harry Cohen, Darryl Zanuck, Louis B. Mayer, Samuel Goldwyn, and Jack Warner) I find fascinating—probably because they had so much control. And the new moguls, Bob and Harvey Weinstein, who are brilliant about predicting trends and tastes—they aren't always right, but more often right than wrong. I admire anyone who is smart and can get a movie made the way they want it to be made. I think the two most exciting, smart producers working today are Christine Vachon and Jennifer Todd. They have completely different tastes—but both are pretty fearless with material and really smart about working with people to get their movies made the right way—and both are extremely supportive of the directors with whom they work.

Interview with Chloe Willow, Ex-producer

In the early nineties, Chloe Willow experienced what it was like to be the "flavor of the month," meaning that she was able to put together a number of projects that became hot property around Hollywood-town. But she was only able to work as a consultant on a number of Gen-X angst-filled youth movies and never got her own projects off the ground. She agreed to share some of her experiences with us—and warned that she doesn't know what went wrong through the whole "long, strange trip."

Chloe's story:

Movies and working in the movies were always my intent . . . but I was turned on by the art films of the fifties and sixties, going out of my way to see Kenneth Anger films at UCLA and stuff like that. I was never mainstream, ever. If it was odd, I liked it. This is probably why my stuff never got off the ground; it didn't reach an audience. There weren't enough eccentrics like me out there to appreciate the projects I had been preparing, or it was culturally the wrong time.

Well, I had found this great writer who lived in upstate New York. I read his manuscript for a company that I was doing coverage for on the side. His book, which was going to be published by a small press, blew my mind and I contacted him, telling him that I couldn't recommend it for the conservative company that I was working for, but that I had to let him know that I thought the book was one of the greatest things ever written. It was my Holden Caufield fantasy come true. Remember how Holden talks about how great it would be if you could call the author of a book that you just read and thoroughly enjoyed and tell them that it was fucking great? Well, that's what I did and I and the author soon became friends. We discovered that we both liked the same very strange things. One thing lead to another and I found out that he had three or four other projects that were equally as great and I decided to call myself a producer and begin to shop him around. He already had a New York literary agent, so I wasn't starting from scratch.

The process was fun. I called all the stars that had deals either as actors or directors and managed to get myself represented by William Morris as a working producer. The writer had some buzz coming out of New York, so the meetings came easier and easier. It looked like everything was going right, and I was able to get three of his projects optioned. Things were looking good. There was no money in it for me until the movie got made, but he made his option money. This went on for a couple of years until none of his projects got made, I was broke, and he wouldn't give me a percentage of any of the money he made off of my pitches of his work.

I was naïve and stupid and didn't get anything in writing. I wasn't even in love with him or anything. I just liked the idea of being passionate about a story and a lead character and getting around town making friends in pitch meetings—which I did, but nothing ever got made, hence I wasn't a very good producer.

Yeah, well, obviously, the subject matter that I had chosen was a bit obscure and odd, and to tell you the truth, I wasn't even thinking about the competition. I was just doing what I wanted to do, pitching movie ideas that I wanted to see . . . it's just that no one else wanted to see these same ideas—a few people got interested—but it wasn't enough to gather support throughout the community so that the writer became so hot that we were all able to retire to the South of France.

The original Andy Warhol films, John Waters, and, as mentioned, Kenneth Anger, those are my mentors . . . anyone who has done experimental film that has crossed over into the fabric of normal popular culture in some way, form, or fashion, is my hero. As far as inspiration, I'd have to say that going to museums of modern art—art from 1950 or so on, that's inspiration. Watching strange old B-movies and what they were attempting to do, even Ed Wood, that stuff is great. That's where I get my inspiration—oh and in the "Chance Meeting" section of the classifieds. There are a million stories to be told within those little snippets of life. That subplot was touched upon in the indie film *Ghost World*, which I loved, where the leads tried to trick Steve Buscemi into a date situation. Love that idea as a premise for a movie. I just think there's something there . . . see how my mind works? It's too weird.

CHAPTER 8

PRODUCTION AND CREW

CREATIVE CAREERS IN HOLLYWOOD

STATUS

DURABILITY: Keeper.

LENGTH OF STAY: Project by project for your entire career.

FOOD-CHAIN VALUE: Mid-level.

UPWARD MOBILITY: None, really.

VACATION: During hiatus.

SALARY: Regular black coffee, bottled water, or beer.

HOW EASY IT IS TO GET THIS JOB: On a scale of 1 to 10 (1 being the easiest), 3.

PREREQUISITES: Having a good work ethic. Being a team player. Being alert and consistent all the time. Having the desire to live a relatively normal life in a dream industry.

All dream the dream of a life devoted to making dreams real.
—Dialogue from Good Morning, Babylon

This chapter is dedicated to all those souls who wish to work in jobs that are referred to as "below the line"—that is, most of the people who are not seen in front of the camera and are not executives. They are behind the scenes, literally, on a movie set. It takes a group of people to make a movie. Indie film director Kevin Smith, guest-speaking at Northwestern University, said that his audience would never see the credit "A Film By" on any of his films, for he knows that he needs his crew, he appreciates his crew, and without them, he couldn't make his movies. Smith is right—the director needs his crew. In this chapter, we are going to explore films that honor those folks who put in long hours on the set.

IN PRODUCTION

There are a number of key positions needed for every film shoot, no matter what the budget of the film. The key to a well-maintained set is to have a good crew filled with people who are organized and coordinated. In most cases, working on a film crew requires many more hours than the normal nine-to-five job. With various temperaments, egos, technical snafus, and acts of God, which can all play into the picture, a movie set is a hotbed of excitement, uncertainty, and creativity.

Working on a set is very different from working behind a desk. The industry can be divided into those who choose to work in an executive capacity, and those who like to be "in production." Here is a brief list of the positions found on a film set:

- **Director**—individual in charge of the production, controls the set.
- **Assistant Director**—assists the director and handles technical supervision.
- **Director of Photography**—works with the director to get the right shot, to realize director's vision.
- **Camera Operator**—physically handles the camera for the DP.
- **Lighting Director**—works with the DP to get the right lighting for the shot.
- **Unit Production Manager**—person in charge of the crew.
- **Location Manager**—scouts shooting locations.
- **Gaffer**—the chief electrician, supervises all lighting on the set.
- **Grip**—the physical laborer of the team. Key grip, best boy grip, dolly grip, company grip—these are all titles for the specific items these individuals move, erect, and strike on the set.
- **Property Master**—person responsible for the rental, purchase, and construction of all props required by the script or story.
- **Sound Mixer**—reports to the director regarding all sound on the production set, works with the **Boom Operator** who maneuvers the long microphones.
- **Wardrobe, Makeup, Hairstylist**—all are needed on the set to assist the cast in proper look and attire for the story.
- **Production Designer, Construction Coordinator, Graphic Artists**—all work closely with the director and DP to insure that the sets are well executed.
- **Craft Services**—provide on-set snacks and drinks, work with catering service for meals for the crew.
- **Script Supervisor**—aids the director in the continuity of each scene.
- **Production Coordinator**—liaison between UPM, assistant director, crew, and actors.
- **Production Accountant**—handles the money for the shoot.

- **Production Secretary**—works with coordinator and performs overall administrative duties.
- **Production Assistant**—this is the entry-level gopher job.

This is a basic crew list. Many of these positions can be combined or eliminated, depending on the budget of the project.

BELOW-THE-LINE STORIES

As stated above, there are a number of films that illustrate the intensity of working on the set. We'll begin with a look at the overall crew through four films made in the late twentieth century.

"Good Morning, Babylon," Said Mr. Chaplin

Two films produced within five years of one another capture Hollywood in her childhood years. These two films, *Good Morning, Babylon* and *Chaplin*, give insight into what it was like to work in the industry at that time. Both of them emphasize the joy and glory of working in Hollywood. The year is 1915. Let's back up for a moment, to understand the importance of that indelible year in the history of Hollywood.

Good Morning, Babylon (Vestron, 1987)

It was in 1915 that director D. W. Griffith made the first American film of monumental significance, *Birth of a Nation,* a Civil War epic. This film sets new standards for the nascent cinematic art. It features Griffith's advanced montage techniques and the use of varied shot sizes (the first close-up and fade-out shots are seen here.) Griffith recognized movies as art. *Birth of a Nation* is followed directly by *Intolerance* in 1916. This equally monumental film is comprised of four complete stories illustrating the eternal problem of human intolerance through examples of different cultures and eras. It is with *Good Morning, Babylon* that one can begin to understand the making of this epic and how triumphant this work was in early Hollywood during the years of 1915–16.

"Once upon a time there were Andrea, Nicola, and D. W. Griffith . . . An American Fable by Paolo & Vittorio Taviani" proclaimed this movie's press ad, which appeared in print media in the summer of 1987. This Italian-American coproduction has been described as a fairy tale, an odyssey, and a Griffithian melodrama. It is all three. Spanning the time between 1910 and 1917, *Good Morning, Babylon* reconstructs life in early Hollywood. The result is a film that illustrates the union between the ancient arts of the Old World and the new technology of the movies. The story is based on two actual occurrences in the lives of the lead characters of Nicola and Andrea Bonnano.

The Bonnano brothers are forced to leave their work of restoring cathedrals in Italy when their father's business fails. He is a master builder who has taught his sons the art of carpentry. Each of the boys possesses "hands of gold." Their journey to the New World takes them first to New York City, where they marvel at the skyscrapers. Hitching a train westward, they soon discover the realities of the harsh Western terrain. In San Francisco, they find work at the 1915 San Francisco Exposition's Italian Pavilion. They build a Tower of Jewels exhibit, and it is the talk of the Exposition (actual event #1). Actual event #2 is the fact that Griffith hears of the grand Tower of Jewels and calls for the pair to join him on his set of *Intolerance*. The brothers find themselves employed as set designers for the famous Babylonian sequence.

In the newest of the new American cities, the naturally surreal environment of the early silent-film industry of 1915 Hollywood is carefully revealed, as the lives of these two builder brothers unfold. The film shows the camaraderie of free spirits who were drawn to the magic of this new art form, all born of the same cloth, "all dream the dream of a life devoted to making dreams real." They ride the Hollywood Red Car, Los Angeles's first attempt at mass transit. It was the heyday of pie-throwing, all-night parties, and instant romance. Within this new place, so clean, so chaotic, so filled with optimism, this fresh community creates a new language of film. And the brothers discover love.

Edna and Mabel ("We can't be extras all of our lives!") are two lovely day players. Nicola and Edna and Andrea and Mabel soon declare their love for one another and the quartet's lives unfold as work goes on under the

Californian-Tuscan sun. Babies are born, and World War I was underway. Vincent Spano and Joaquim de Almeida are the lead characters, and Greta Scacchi and Désirée Nosbusch play their mates. Charles Dance portrays Griffith with magnificent grace.

The scene that is not to be missed by anyone who has a desire to work in the industry is at the couples' shared wedding ceremony. Their aging father and other guests join the brothers and their new wives. Griffith himself, who was truly a pioneer of filmmaking, gives a speech that defends moviemaking as a legitimate art. In real life, Griffith was a proud, witty visionary who always kept a very moral set. He says in his speech that the cinema is the work of many artisans — "It is a collective work. Lots of people contribute to film." His toast continues to evoke the power of the cinema and its ability to move people and unite them . . . (it is) "an art that holds out the promise of world peace and understanding." This scene is a visual tour de force, operatic in style as it takes place with the used Babylonian set as a background for the wedding ceremony. The other memorable scenes are those that reveal the actual art of silent moviemaking. Numerous shots explain how primitive production techniques took place with natural sunlight pouring across a golden meadow or through an aperture onto the studio stage to spotlight dancers dressed as mythological figures.

Chaplin (Tristar, 1992)

Charles Chaplin worked in director Mack Sennett's Keystone Studios, right down the road from Griffith's studios. The producers of *Chaplin* had as much trouble recreating post-WWI Hollywood in 1990 L.A. as the producers of *Good Morning, Babylon* did a few years earlier. Thus, Sennett's Keystone Studios, Chaplin's studios, and the location of the Hollywood sign had to be reconstructed for the sole purpose of the movie's production. To see how Hollywood looked three generations ago, it was necessary to drive sixty miles north of L.A. to Fillmore, California. It was Fillmore's orange groves and nearby hills that doubled for that rural look of Hollywood circa 1915.

Director Richard Attenborough placed prime importance on being true to the time period. To recreate the look of the early films, Sven Nykvist, the

film's director of photography, utilized as few electric lights as possible, rely-
ing, as did the early filmmakers, on muslin-filtered light only. The scenes
that are shot on the muslin-draped open-air stages do have a magical, whimsical, almost ethereal feel to them, taking the audience back to the wonderment of that time. When Chaplin walks alone onto the empty stage, as
the cloth billows in the light, the visual effect is memorable . . . again, the
Tuscan-Californian light shines soft. In addition, Sand Canyon in the
Angeles National Forest, also located about a half-hour northeast of L.A.,
doubled as the hills surrounding the Hollywood sign.

Today, the Chaplin studios (1416 North LaBrea, just south of Sunset
Boulevard, the location of a number of different production and music companies over the years) capture the essence of what Hollywood must have been
like in these times. There stand the small, Tudor-style bungalows. They were
built in this manner in order not to offend the neighbors. Other studios during this time are barnlike and factory-built. Chaplin shot his films on this lot
and remained there until he moved to the United Artists lot many years later.

Chaplin offers a look at the crews and production processes of this time.
These crews had to control wind and other weather elements. Director
Attenborough achieves a healthy balance of the story of Chaplin's life and
the world he lived in. This world of Hollywood soundstages and back lots
comes alive for modern-day audiences a century after the real-life events
took place.

Day of the Locust (Paramount, 1975)

Long regarded within literary circles as *the* novel about Hollywood, this
Nathanael West's story is a dark slice of Hollywood life as if seen under a
magnifying glass through the eyes of one cynical art director Tod Hackett
(William Atherton). The studio atmosphere of the thirties is reproduced as
Tod works on the sets of productions and falls in love with an actress who
lives in his apartment building, the San Bernadino Apartments, which he
describes as "early earthquake."

Scenes of Tod on the set are abundant, but the real reason to watch this
movie is to see West's Depression-era Hollywood peppered with all of the lost

souls that surround Tod. West has an affinity for garish, grotesque bodies: dwarves, prostitutes, clowns, cowboys, and everyone else who has come to California to seek their fame in the movies. What began at the turn of the twentieth century and is seen in the two previous movies is alive and well and thriving in *Day of the Locust*—as alive as the despair and unrealized dreams.

Living in Oblivion (Sony Pictures Classics, 1995)

Fast forward to 1995 for the next film to feature insight into the crew and all of its complexities to an independent film titled *Living in Oblivion* directed by Tom DiCillo. The film is set up as a series of episodes featuring Nick Reve (Steve Buscemi), the director, and Chad Palomino (James LeGros) as the lead actor. In each scene that Nick tries to complete, a different problem plagues the shot. In one take, the boom mike slips into the frame; in another, a noise outside is picked up on the soundtrack. Finally, the only genuinely good performance is not filmed because the director of photography is in the bathroom, throwing up the spoiled milk from the craft-service table.

These opening scenes turn out to be Nick's nightmares; however, once he does begin real production, we are introduced to Chad, a pretentious young actor who screws up every take by trying to better his own image. All of this is exacerbated by the fact that the two lead actors slept together the night before and are talking about one another behind each other's backs. A physical brawl involving most of the members of the cast and crew breaks out; but then it is revealed that this is really the lead actress's nightmare.

The final scene is Nick's shooting of an actual dream sequence. In all of the above scenes, the dynamic between the director and his crew must be observed. Nick feels as if nothing will go right on the set and that no matter what he does, the movie will turn out bad. His confidence is waning. Through his unsteady thoughts, we get to see how a movie set works.

There is a definite hierarchy established on the set, as only the actors, director of photography, and assistant director talk to Nick. The actors want more screen time, the DP wants to change the shot, and the AD just kisses Nick's ass. Nick has to be involved in solving every problem, from tweaking a light to altering the script. Learn from Nick's actions. Directing a film, working on set, dealing with fellow crewmembers can be a harrowing experience. No matter how smooth a movie set is, there is inevitably going to be some miscommunication. Tom DiCillo's *Living in Oblivion* is an accurate portrayal of a movie set—major budget or independent, there are always going to be some bumps in the road.

The film is horribly credible, the ultimate low-budget film about making a low-budget film. It's been called "*The Bad and the Beautiful* of indies."

Speaking of Bumps in the Road . . .

The next movie takes a look at stuntpeople, those brave men and women who act as stand-ins for stars when situations in the script call for dangerous acts. What follow are discussions of four films that explore this topic, and all but one are murky in execution. This is a wild breed of folk. Let's take a look.

The Stunt Man (Twentieth Century Fox, 1980)

The Stunt Man is a perfect example of a movie-being-filmed-within-a-movie. The frequent use of scenes being shot for a movie, within the movie, demands that the audience follow along closely. Okay, it's a common trick to open a movie with a shot from a set or utilize a scene that is being shot on location, but soon someone yells "cut," and the movie you are watching in present time begins. In *The Stunt Man*, the movie action moves in and out of reality and fantasy.

The situation is a perfect hideout for fugitive Cameron (Steve Railsback, generally known as a TV actor), who is both aided and endangered by a maniacal director portrayed nastily enough by Peter O'Toole. The term "sudden death" takes on a new meaning as this new stuntman endures the perilous feats he is asked to perform on the World War I epic being shot. The life of a stuntman is revealed as he earns his living and eventually falls in love with the leading lady. *The Stunt Man* is an early example of a meta-movie, a movie that mocks the making of movies—and it does it very well. This film itself took nine years to shoot—in part because the director, Richard Rush, suffered a heart attack during production.

Honorable Mentions

Hollywood Thrillmakers (Lippert, 1954)

The Last Movie (Universal, 1971)

The Great Waldo Pepper (Universal, 1975)

There are three remaining films to view if you are interested in working in this industry as a stuntman. They are: *Hollywood Thrillmakers,* a generic look at an average stuntman's life that utilizes stock footage from the vaults of Hollywood; *The Last Movie,* starring Dennis Hopper as a stuntman veteran, and *The Great Waldo Pepper,* a Robert Redford-as-stuntman vehicle. This is an extremely dangerous, thrill-seeking profession that features self-appointed heroes in constant peril. It is, quite simply, only for a chosen few.

HONORING THE CREDIT CRAWL

Those skilled individuals who work "in production" and as part of the crew are often referred to as "below-the-line" expenses, as mentioned at the beginning of this chapter. "Below-the-line" are labor and technical expenses, such as set construction, camera equipment, film stock, and developing and printing. (As opposed to "above-the-line" costs, which cover all of the major elements of a film, such as the writer, director, actor, producer, and the script and story development costs.)

The lesson to be learned is that each one of these crew people has dedication and determination to produce entertainment. The long hours on the set, behind a camera, or as a wardrobe assistant, prove that the people who choose this as their life's work are devoted to working as part of a team of journeymen within the dream factory known as Hollywood. These are people who make up the seemingly endless lists of names you see in the credits of every film.

Keep in mind that in Hollywood you must stay for the credits out of respect for every single person who helped to make that film. If you leave immediately when the movie is over, you'll betray a long-standing bond between you and your fellow workers in the industry. You'll also be frowned upon by others in the audience, *especially* if you have just attended an industry screening where many of the individuals who worked on the film are actually present in the audience. So, don't leave early. Stay and be thankful that you are part of a group of people who are privileged to be part of the moviemaking machinery.

IN REAL LIFE

Being in the credits of any production is thrilling. Both of these workers stay glued to the screen to see their name in lights.

Real Life Journeymen

The first, Todd G. Todd, has been a PA, a gopher, a runner, an extra, a commercial director, a short-film director, a stage manager, an actor, a sound editor, and a talk-show producer. He is currently working as a producer-director of interstitial programming.

On movie influences:

No movie really influenced me to get into show business; in fact, I often wonder how I ended up in show business. I don't think I chose it as a career, and with the many extended bouts of unemployment, I wonder if it chose me.

On expertise:

I have many areas of expertise, from producing to sound editing to anything you can imagine in between (legal or otherwise). When I was starting out, I wasn't picky, so I ended up with a very well-rounded skill set and some incredibly useful sleazy friends; overall it left me with enough knowledge to understand what was being asked of me and a Rolodex of people to call to find out how to do it.

On competition:

Competition doesn't really play into it. I was once fired for eating popcorn, luckily, this was early on so I realized that there was no rhyme or reason to why some fail and some succeed at the hands of the heathens who run Hollywood.

On why he does what he does:

I really don't know why I still do it. You tell yourself it's for the money, but when you do the math for the amount of hours you work and the BS you endure, you can make way more money at The Gap, and from what I'm told, nothing is more rewarding than a properly folded pair of chinos. I do think that there is a hurry-up-and-wait thing going on, but I would not call crews lazy. They are some of the most dedicated, hardworking people I've ever seen, and as long as you are talking about goofing off, you can't imagine the lengths some of these guys will go to to put gaffer's tape on your back or put a laundry pin on the tail of your shirt.

On a favorite shoot:

My favorite shoot was in the beginning of my career. I was doing sound for a pilot, and had to mike a six-foot-tall Penthouse Pet of the Year who wore nothing but heels, a wedding veil, and a string of pearls. The room we were shooting in had a low ceiling, so I had to lie down on the floor in between her legs and mike her from there. What this young lady lacked in acting skills she surely made up for in accessorizing.

Our second artisan is Timmy G., who has mastered the art of editing in his four-year career in Hollywood since graduating from college with a B.A. in Film.

On movie influences:

Star Wars and *The Empire Strikes Back* were my primary influences because of their ability to tell an amazing far-reaching story, and to be watchable again and again, and to affect other people all over the world—that's a pretty tough act to follow. Nevertheless, sadly, just like every other child born in the seventies, I thought that I could follow it. But, as it turns out, my real favorite films came to me much later—during college, when I discovered *The 400 Blows*. Most of my favorite films nowadays actually do not come from Hollywood, or, for that matter, the United States.

On new ways to edit:

There are always different ways to tell stories—always a different way to edit a scene—always a different take to use. The question you always have to ask is— did I accomplish the goals of the writer and director while ensuring I did the best job in the process? As for what audiences like . . . I don't think anyone can accurately predict what audiences will like or won't like. When I was in film school, we read a book by William Goldman where he claimed that *nobody knows anything*—and I think that sums up most of moviemaking. There is a difference between a good and a bad film, but whether or not audiences want to see something—that is entirely impossible to predict, unless you are J. K. Rowling.

On competition:

With the editing world, editors are rather bonding together than competing. Editors go up for the same jobs and some get them, and some don't—but more than likely, there are no hard feelings. Since there are so many television shows and feature films out there, a good editor or assistant editor can find work. And, unlike in the world of development, an editor generally has to be damn good at their job to continue to be employed.

What makes me good at my job is my thoroughness, organization, passion for technology and storytelling, sense of humor, ability to socialize well, and ability to sense the inherent politics that go back and forth on a production. This is crucial so you don't make a fool out of yourself. But I am certainly not the only one who knows how to do this. Editors are the unsung heroes of moviemaking.

CHAPTER 9

STUDIO EXECUTIVE

CREATIVE CAREERS IN HOLLYWOOD

STATUS

DURABILITY: Keeper.

LENGTH OF STAY: Two to five years, or as long as you can handle it.

FOOD-CHAIN VALUE: Senior level.

UPWARD MOBILITY: It's the top; it's only downhill or a development deal from here.

DESIRABILITY FACTOR: Supreme.

VACATION: The company's annual retreat in Palm Desert or white-water rafting with peers in the Pacific Northwest.

SALARY: Triple-shot Venti Iced Blended Special.

HOW EASY IT IS TO GET THIS JOB: On a scale of 1 to 10 (1 being the easiest), 11.

PREREQUISITES: Starting out in the mail room of a studio or major agency. Being born into a Hollywood family. Having made a fortune in some other industry. Being a really good dresser and looking like a Ken doll. Having some golfing buddies in high places helps too.

Are there still Sammys in Hollywood? . . . the Sammy-drive is still to be found everywhere in America, in every field of endeavor and among every racial group. It will survive as long as money and prestige and power are ends in themselves, running wild, unharnessed from usefulness.
—Budd Schulberg, 1952 introduction to his 1941 novel What Makes Sammy Run?

The suits. The front office. The top brass. These are the big shots, the guys (mostly) who can greenlight any project, the guys with all the power. The job of an executive at a studio is highly coveted. This position is intense, not always rewarding, and chock-full of surprises. Power is the key word here. A studio exec has the power to choose the content and direction of his studio. He holds in his hands the destinies of many individuals. This is a radically demanding position that requires long hours of meetings, phone exchanges, tons of industry luncheons, and the painstaking babysitting of egos. This is the individual who puts the fires out. Many studio execs are not unfamiliar with the use of Prozac or the need to visit a therapist, whether massage or psychological, on a twice-weekly basis.

There has been a shift from movie mogul (having the creative *and* the business sense) to financial wizard (having the business sense and very little creative ability) in this powerful office as the years have progressed. Ideally, this position should be a blending of the financial savvy with the creative taste, in order to successfully choose the projects and producers who will provide solid box-office performance. The studio exec oversees all of the other departments within the studio. He must be well aware of his company's strategy and what his competition is doing around town as well as globally. In addition, he must have his fingers on the pulse of the industry and be well aware of every important and upcoming agent, star, producer, director, and every other hot commodity in order to woo the popular ones to be on his side and do movies with his studio.

On the personal side, studio executives have the best offices; most of them are properly feng shui-ed and decked out in either a Southwestern, Japanese, or Hi-tech motif. They drive sleek, fast, and expensive luxury cars, live in the Palisades, Bel Air, or Brentwood, and shop at Barneys or Fred Segal. They control popular culture. They are the modern-day gods of the silver screen. F. Scott Fitzgerald chose Monroe Stahr, the lead character of his last novel, *The Last Tycoon,* to be the embodiment of the quintessential studio executive. Stahr exists as if he were the savior, the new creative head of the studio, the one who will set all standards for the future. Many of the early Hollywood moguls were, in fact, seen in just that way, as rulers who had a view from the Heavens and built their own empire to reinvent themselves and America at the same time. We will look at the various films that have portrayed successful movie heads, but before that, let's review the top positions of power at a studio.

EXECUTIVE ROSTER

Modern-day moguls, barons, and merchant princes (and princesses)—here's a breakdown of studio power.

Studio Chief

This position is usually also the CEO and chairperson of the studio. He is the representative to the community and industry, the spokesperson when called upon to promote or defend his studio's creative endeavors. This person has the final say in every decision involving the studio. Very often, though it is not mandatory, he holds an MBA from a very prestigious university.

President of Worldwide Production

It is generally up to this person what projects the studio produces. He is in charge of the studio's strategy, including the selection, development, production, and distribution of its projects. He also plans the film-release patterns for the spring, summer blockbuster, and holiday films. He is the heart of the studio and again, more often than not, is in possession of a higher degree in business or law.

Vice President of Production

Reports to the president of worldwide production and the studio chief. Maintains good relationships with the community. Must be in touch with all of the agents and stars, and have both the creative and financial backgrounds. Does the extra schmoozing that both the production president and studio chief do not have time for. May or may not hold an advanced degree, but most certainly a Bachelor of Arts at least.

The percentage of individuals who reach the level of studio chief is small, yet it is the position that many are working toward within the Hollywood Food Chain. There are four movies (and one honorable mention) that deal specifically with the role of studio executive. One of the first films about studio executives is *The Last Tycoon*, based on Fitzgerald's novel. Fitzgerald modeled the character of Stahr on the mogul Irving Thalberg, one of the most legendary individuals in the history of film. Thalberg demanded his employees' complete dedication to their work and to the studio. He would call them at odd times, insist that they show up immediately, and milk them for ideas. His studio system was directed toward one end—perfection. He worked compulsively, intuitively, and restlessly, and he had three rules of success:

- Never take any one man's opinion as final
- Never take your own opinion as final
- Never expect anyone to help you but yourself

That last rule is one that should be taken to heart by anybody who wants to take on a creative career in Hollywood—never expect anyone to help you but yourself—this is the truth. At any rate, Thalberg was and remains a most remarkable figure, and in many ways, he represents Hollywood's soul. His premature death at thirty-seven would always keep him wrapped in nostalgia and romance, a picture of perfection.

SCREEN TYCOONS

Here now are the films that feature movie executives in their natural habitat.

The Last Tycoon (Paramount, 1976)

For anyone sincerely interested in running a studio, this film is a must-see. Sadly, at the end of his life, as he was writing this unfinished novel, F. Scott Fitzgerald had become a desperate has-been. He became as disposable as the silent-movie stars in *Sunset Boulevard*. He had been a voice of a generation, only that generation had grown up. Still, this late novel, and the movie by same name, captures the loneliness and the dedication of the front-office men during the thirties.

This tale of studio politics was directed by Elia Kazan and scripted by writer Harold Pinter. The movie did not find its audience, mostly because the plot is slow moving and not executed in a linear manner. The narrative reflects Stahr's thoughts as he pursues a beautiful young woman who reminds him of his now-deceased wife. When the young starlet rejects him, Stahr suffers a nervous breakdown. His thought sequences make the narrative difficult to follow.

Robert De Niro, however, captures Stahr's essence brilliantly. He is impeccable in this role. He is precise, determined, and persistent. If one wishes to be a studio head, one could learn a great deal from De Niro's performance here. The first scene is a walk through the Paramount lot as Stahr tells a writer the truth about his script. "Writers are children, they are not equipped for authority," says Stahr. He feels the need to be in control of every element of his studio, including the writers. He expresses this thought through the film's next scene, a scene that completely captures the magic of the movies.

Donald Pleasence plays a writer who is having difficulty writing melodrama in a scenario Stahr has assigned him to. Stahr sits down at his executive desk and begins to tell him a story. "Suppose you're in your office . . . a pretty stenographer that you've seen before comes into the room . . ." The essence of the short story that Stahr tells is that the stenographer then opens her purse and dumps its contents on the table—two dimes, a nickel, and a matchbox. She leaves the nickel on the desk, puts the two dimes back in her purse, and takes her black gloves and burns them in the fireplace. Seconds later, she receives a phone call, answers it, and deliberately says: "I've never owned a pair of black gloves in my life."

Stahr finishes his story leaving the writer wanting more. The writer says, "Go on . . . what happens?" Stahr answers quite matter-of-factly, "I don't know, I was just making pictures."

The writer feels like he has been shortchanged. "What was the nickel for?"

Stahr turns to his secretary and asks her about the nickel. She confidently reports: "The nickel was for the movies."

"What in the hell do you pay me for?" the writer demands. "I don't understand the damn stuff."

"You will," Stahr says grinning, "or you wouldn't have asked about the nickel."

This scene shows that the writer character, like the great American writers who went to Hollywood in the thirties, has to learn a whole new "unwriterly" way of writing. Stahr proves that he can spin a tale just by "talking pictures," while the writer, used to complex word structures, has to be taught to convey story through instantaneous images. The nickel question? A nickel, of course, was the cost of movie admission in the thirties, and symbolizes, within the story, the details good writers need to pay attention to. This scene is repeated later, at the end of this movie adaptation of an unfinished novel; the original story does not have an ending, and therefore, this is a movie without an ending.

"I don't know, I was just making pictures," Stahr admits innocently. And make pictures he did. Stahr is a young man who possesses wonderment for life and for reinventing life on-screen. He is a man who lives and breathes moviemaking. His all-work-no-play lifestyle makes Stahr a lonely man. Originally, this movie was labeled as a romance. *The Last Tycoon*'s main source of conflict arrives when Stahr's love for a pretty starlet is not reciprocated. He realizes how much is missing from his life because he does not have love. This is the price he must pay to be the most successful executive of his time. His role in this newly created movie industry, along with his sheer determination, results in a life of loneliness, shallow and empty. In many ways, success today demands the same sacrifices. It is difficult to keep a personal life together when the demands of the studio persist 24-7.

When Stahr spends time on the studio's back lot, it's as if he has retreated to his own land of make-believe. It is here, and only here, that Stahr can make life real for himself. Functioning in the real world can be difficult when one is working on such a fast track and in such isolated circumstances. In this case, it is lonely at the top.

Honorable Mention: Irving Thalberg

Thalberg's influence on Hollywood and the film industry was extensive.

Man of a Thousand Faces (Universal, 1957)

In addition to *The Last Tycoon*, there is one other film that features the character of Irving Thalberg in the plot, and that is *Man of a Thousand Faces*, starring James Cagney as character actor Lon Chaney. Interestingly enough, actor-turned-studio-executive himself, Robert Evans, portrays Thalberg and is featured sporadically throughout the narrative as the studio exec that continuously supports Chaney's unique career moves in choosing acting roles that required contortionist actions and heavy makeup. This movie features Thalberg in his role as studio mentor, showing his personal concern for the careers of his actors, both personally and professionally.

The Bad and the Beautiful (MGM, 1952)

If you want corroboration for the myth that those at the top are evil puppets, check out vein-popping, jaw-thrusting Kirk Douglas in this glossy black-and-white film that looks like an animated gelatin-print photograph. In grand fifties fashion, this movie rolls out onto the screen, lush in great Hollywood grandeur. Featured here are Dick Powell as James Lee Bartlow, a screenwriter ("You can always tell a successful author by the cashmere jacket"), Lana Turner as actress Georgia Lorrison, and Barry Sullivan as director Fred Amiel. All three tell their stories of what it was like to work with the ruthless and cunning megalomaniacal producer Jonathan Shields (Douglas) and how he used them as his stepping stones to get to the top in Hollywood. This film features great acting and had Academy Awards nominations for practically everything. It is truly one of the best portraits of Hollywood in the fifties, and most of its content remains true today.

The Bad and the Beautiful is definitely one of the best films about moviemaking. Director Vincente Minnelli, producer John Houseman, and writer Charles Schnee worked meticulously to create a film that would mirror real life. Jonathan Shields is not a popular man. He actually hires mourners to attend his father's funeral to make it look like someone cared. Shields admits his father was despicable, but he turns his father's film business around and begins to experience great success. He gives those thinking of working in Hollywood some of the best advice — "If you dream, dream big." (See *A Star Is Born* discussion in chapter 1.) He also reminds us of the following: "The best movies are made by people working together who hate their guts." Shields is not kind. Each of the three recalls their escapades with him through flashbacks. Georgia abandons drink and despair after falling in love with him. Bartlow gets the encouragement from Shields to write commercially (although, at the same time, he loses his wife due to Shields's meddling in his life). Director Amiel feels used by Shields, yet thanks to his association with Shields, Amiel's career skyrockets. So how bad is Shields if he has driven each one of these individuals to do their best work?

Not bad at all. In fact, he is probably one of the most effective producers in the history of films about working in Hollywood. In the finale of this movie, all three join forces to make a new film for Shields, who must engineer a new comeback for himself. Another reinvention is about to get underway. One can learn from Shields. You must be tough. You must be steadfast and strong and you must remain true to your vision to make it work. Yeah, the people working with you will hate your guts but it's the only way to get things done in Hollywood. Know what you want. Go out and get it. Dream big. Shields is an excellent character to study, to observe, to emulate if you want to be successful within the high ranks of Hollywood. Shields is not a gentleman like Stahr. Shields wants complete control. He is opportunistic and acts upon it. However, as evil as Shields is made out to be, he pales in comparison to studio heads of the future.

The Player (Fine Line, 1992)

As we look at the films that provide the most striking depictions of Hollywood careers, three films seem to pop up over and over again. They are *Swimming with Sharks, The Player,* and *Sunset Boulevard. Sharks* and *Player* are essential movies for those interested in becoming a studio chief, because they feature the same position Monroe Stahr and Jonathan Shields so righteously filled. Fast forward to the nineties. Stahr and Shields have morphed into Griffin Mill and Buddy Ackerman, respectively. One has a little more restraint than the other does; yet they both pack a powerful punch.

Like Stahr, Griffin Mill (Tim Robbins in *The Player*), has class and style. He dresses impeccably. His demeanor is professional, calm, collected. Mill conducts his pitch meetings, his celebrity-sighting and celebrity-sharing lunches, and his typical Hollywood weekend getaways (at Two Bunch Palms in Palm Desert, no less) in the usual way of a nineties studio exec. He drives the proper SUV and orders bottled water wherever he goes. He is the picture of studio success, a grown boy (Robbins's youthful face fits perfectly here) with all of his toys—actors, directors, writers, budgets, the studio's reputation—stacked up in place for him to manipulate accordingly. He seems to be in control until a rival from another studio, Larry Levy (Peter Gallagher), gradually threatens his position as *the* golden boy creative executive at the studio. Mill is aware of Levy's growing popularity and slick escapades to take over his position; however, Mill has another threat looming on the horizon.

Postcards. Threatening postcards have been arriving in Mill's in-box, stuck behind his windshield wipers, and delivered to the studio lot, seemingly from a downtrodden writer who wants revenge on Mill for not listening to and buying his story idea pitches. Mill consults his date books and pinpoints the writer he believes has been sending the threatening missives. After meeting the writer at a Pasadena theater, Mill loses his temper and kills the writer. For the remainder of the movie, Mill works under more pressure than he could have ever imagined. As Levy moves in and up the studio's ranks and Mill's conscience gains on him, Mill plays out his role as a studio exec. He courts and begins an affair with the dead writer's wife . . . and he receives more communications from the rejected writer. Did he kill the wrong man? Is his career still intact? Interactions with the police become part of his daily life, yet the final scene shows us that his storybook existence continues as he drives up to a beautiful home and a beautiful wife, and everything is hunky-dory. This studio exec is so good he even gets away with murder—quite literally. Watch this film for confidence-building. If you have the confident demeanor of one Griffin Mill, pre- and post-murder, you will surely make it in Hollywood. No doubt.

Swimming with Sharks (Trimark, 1994)

Kevin Spacey's Buddy Ackerman is every Hollywood wannabe's nightmare. He is the senior executive vice president of production at Keystone Pictures and is notorious with the industry for insulting and humiliating his assistants. He could be your first boss in the business. If you survive your apprenticeship with Mr. Ackerman, you're guaranteed to succeed.

Swimming with Sharks tells the tale of beleaguered assistant Guy and high-powered exec Buddy. "Shut up, listen, and learn," Buddy repeatedly shouts at Guy as he lobs paper clips at him and continually abuses him with statements like "If you were my toilet, I wouldn't bother flushing it" and "My bathmat means more to me than you!" Buddy forces Guy to place an urgent call to somebody who is white-water-rafting with Tom Cruise and berates him for bringing an Equal when a Sweet'N Low is requested. "You have no brain. No judgment calls are necessary. What you think means nothing.

What you feel means nothing. You are here for me. You are here to protect my interests and to serve my needs. So, while it may look like a little thing to you, when I ask for a packet of Sweet'N Low, that's what I want. And it's your responsibility to see that I get what I want." Buddy assures Guy that his struggle and suffering will be worth it because "this job is very big on payback" and that Rex, his former assistant whom Buddy calls "dogboy," has gone on to be a vice president at Paramount.

Buddy makes bold statements about the work he is doing. He believes he is in a business that develops people's dreams. He lectures Guy. "And learn from this," he says. "If they can't start a meeting without you, well, that's a meeting worth going to, isn't it? And that's the only kind of meeting you should ever concern yourself with." Buddy has perfected a great talent for exploiting others, along with a withering gaze that pierces right through you. He makes Guy track down a blonde in the hall—"West lobby, tube dress, stiletto heels, hurry. Fetch!"—later promising her a part in a picture if she'll have a date with him that will begin at midnight over at his house. He ruthlessly competes with a studio rival who is in a job he covets and has been praised as the maestro of "wham-bam action." Guy's job is to "protect Buddy's interests and serve Buddy's needs" while Buddy teaches the newbie about sabotaging friends, undermining enemies, and loving nobody but himself. Buddy is Shields on steroids. He is a very real depiction of the type of executive animal that oftentimes resides in the offices of Hollywood.

The movie then takes a turn as Guy gets his revenge on Buddy, taking him hostage in Buddy's own house and torturing him physically and berating him verbally. Buddy does his best to get Guy to stop the pain, explaining: "Life is not a movie. Good guys lose, everybody lies, and love . . . does not conquer all." In the climatic scene, we learn that Buddy's wife was raped and shot dead, and that his bitterness and cynicism stem from that loss. His pain and anger had been transformed into a blind drive. Guy and Buddy eventually see eye to eye, and Guy is promoted at Keystone. (To understand Guy's point of view on this situation, see chapter 3.)

Buddy Ackerman is a pure player. He is one of the power players in Hollywood. He lashes out, berates, castrates, and castigates his workforce.

He also gets things done. He is one of the handfuls of people who can greenlight and make movies. He is not a servant to anyone, anywhere.

PLAYING AND SWIMMING

These movies feature four very distinct men, occupying more or less the same position in the entertainment industry. Stahr is to Mill as Shields is to Ackerman. As much as the first two are gentlemen, the other two are assholes, yet they all share the determination and persistence needed in these studio executive roles. Powerful and commanding, Stahr remains the perfectionist, Shields the essence of the studio system, Mill the slick murderer, and Ackerman the senior vice president of all assholes. The view of this job gets bleaker and bleaker as the years go by, yet in the fifties, Shields was to his three principals what Ackerman is to his assistant. Mainly, these men struggle to understand human nature and learn how to be discreet and keep secrets; some push the envelope of life when their love is unrequited and some even commit murder. All this as they manage their studios' budgets and negotiate the never-ending questions about the production of art versus what makes boffo box office. Nothing much has changed in the last hundred years. In the case of studio executive, it is quite clearly lonely at the top.

TODAY'S "OLD MEN"

The original moguls, Carl Laemmle, Adolph Zukor, William Fox, Louis B. Mayer, and Benjamin Warner and his sons, were united in a deep spiritual kinship that helped them to reject their pasts and have absolute devotion to their new country. They embarked upon ruthless and complete assimilation in this new world. This energy is similar to that of everyone who ever has or ever will venture out to Hollywood. There are many people from other parts of the world who relocate to Hollywood for a new life. They leave their pasts behind, they start over, and they perhaps take part in a new dream for themselves. They are as much pioneers as those who arrived over a hundred years ago were.

Yes, the studio head of the new millennium is a corporate animal often trained as an accountant or lawyer. He (yes, mostly "he") is guided by the bottom line and usually pays more attention to the commerce side of the balance of art and commerce than the art side. The business side is winning and the result of that can be seen by what makes it to the public's screens.

If you look closely, you'll see the new crop of future studio executives who resemble Stahr, Shields, Mill, and Ackerman. They arrive and blend into the mainstream of Hollywood pop culture year after year. They create dreams, reflect the collective consciousness of the public, and give it back to the public in the form of movies, in the form of global entertainment. Just like to all the others who choose to work in this business—nothing matters to them but the movies.

IN REAL LIFE

A seasoned executive and a voice that's next in line to rule—here are two executive points of view.

Paul Valley, Mogul-in-training

"I'm just an industry source," explains Paul. "There are two kinds of people in Hollywood—people who make movies and servants to the people who make movies. Only a handful of people make pop culture—everyone else is a speedbump to get to the people that matter."

Paul is a creative executive for a producer who has a studio deal. He is a well-educated twenty-five-year-old man who has worked for two of the most powerful (and notorious) producers in the industry today—and has survived. He now works on the Universal lot as a glorified version of a d-girl. His main objective is to find movies—be they in the form of books, magazine articles, pitch ideas, or scripts—and develop them into the best possible movie, a blockbuster movie.

"Agents suck—some of them suck," he explains. "Some are wonderful . . . smart. They realize that we are all in this together. Others are your adversaries, always lying. Everybody lies in Hollywood.

"Nothing can prepare you for working in Hollywood—it is an apprentice type of system." When asked what his plans for the future are, Paul replies: "I would be disappointed if I was not the next great producer. I do have a voice and I am in a place to have my voice be heard. But I am still servicing someone right now. I need to find a project I am passionate about and go out and produce my own picture. By the time I am thirty years old, I will be a successful solo producer." What does Paul think is the best advice for someone setting his or her sights on being a studio exec? "Don't be afraid to start at the bottom with the right attitude, and work hard. Early on, work ethic is more important than intelligence. Drive is what you need. Drive will take you further than intelligence. It's better to seem intelligent than to be intelligent, and you can't tell the difference with most people out here."

Brenda De Atocha, Eternal Executive

"Sometimes it's hard for me to believe that I've been working at either a studio or major production company for over twenty years now," Ms. De Atocha says. She started out as a reader–story analyst at a production company that got behind one of Tom Cruise's first movies. After that success, she remained employed consecutively by three of the eight major studios until finally negotiating a first-look deal and coheading a production company that supplies one of the major studios directly. She's one of the pioneers, a veteran as far as women are concerned in Tinseltown.

"I've always considered myself a pioneer, but I never let the fact that I was a woman hold me back from any position I was up for—nor did I ever think I was offered less than my male cohorts. I have just been working and doing my job during these past twenty years. Yes, in the early to mid-eighties I was probably being paid less than the men who were doing the same job, but that has improved and at no point did I enjoy my job less," Brenda explains, adding that, as many women would agree, the climb up the ladder hoping to break the glass ceiling has not been an easy one. "I guess I just don't really dwell on the fact that an individual is a male or a female, I honestly want to hire and work with the individual who can do the job—and I expect no favors in that respect from anyone else."

Brenda continues to talk about her philosophy behind being a successful woman executive and whether or not she would prefer to work with men over women.

Most definitely I would prefer to work with men over women. Men know how to compete. Men know how to work without the messiness of having feelings involved. Women, on the other hand, are extremely namby-pamby about how they conduct business. They often concentrate on trivial things. I just want to get the work done. I really don't care about anyone's kids when I am in the middle of a major deal. Some might call me cold for a woman, but talking about the well-being of a person's family is for Saturday and evening phone chats. When I'm in the middle of an intense pitch meeting or a deal, I really couldn't care less about the personal side of the people around me.

Brenda continues with advice for young women who are wannabe-moguls:

Don't give up and never think you are inferior to anyone else—male or female. Be your own person. Be who you were meant to be. Remember that line from *The Fountainhead* when Howard Roark is told that he cannot continue to design buildings the way he wants to—in his own unique pattern. His subversive boss asks him, "Do you mean to tell me that you're thinking seriously of building *that way* when and *if* you are an architect?" Roark answers yes, as the boss continues: "My dear fellow, who will let you?" and Roark boldly responds with "That's not the point. The point is, who will stop me?" And that is the way everyone who wants to be in this business must embrace his or her work—young women and men alike. It's not who is going to *let* you do this job, it's who is going to *stop* you—that's who you need to be on the lookout for.

CHAPTER 10

WRITER

CREATIVE CAREERS IN HOLLYWOOD

STATUS

DURABILITY: Keeper.

LENGTH OF STAY: A lifetime.

FOOD-CHAIN VALUE: Lowest on the totem pole.

UPWARD MOBILITY: None, really.

DESIRABILITY FACTOR: High, especially among trust-fund kids, frustrated lawyers, and soccer moms.

VACATION: All the time, it seems.

SALARY: Black coffee.

HOW EASY IT IS TO GET THIS JOB: On a scale of 1 to 10 (1 being the easiest), 1 if you are writing a spec script; 10 if you want to be gainfully employed as a Hollywood writer.

PREREQUISITES: Being moody, dressing in an offbeat manner, knowing the latest street-slang, and being an Ivy League college alumni.

What's all this business of being a writer? It's just putting one word after another.
—Irving Thalberg

Writers in Hollywood have the unique challenge of writing for the screen, writing visually, not just on paper but writing words that will be transformed into images. In order to write for the screen they must have the capacity to envision their characters, story lines, and dialogue as if it was all unfolding in large, horizontal tableaux. Writing of that nature is very different from writing for the page alone. In addition, unlike novelists and fiction writers, screenwriters have some very rigid format requirements. Essentially, they must learn to use words economically. Writers need to successfully transfer their mind's visuals to the page in order for the director and crew to translate those words into screen images. Screenwriters write the blueprint of the movie, and in writing that blueprint, they produce a work called a screenplay.

The writer in Hollywood is a maverick. He is a beacon of information, yet everyone treats him like dirt. He is the most necessary part of the moviemaking process, for without him there is no story. The writer is also the most abused. His words are the very essence of what is appearing on the screen, yet he is rarely consulted or called upon when the director and crew decide to take liberties with or change his script. Writing a screenplay is like giving birth and immediately giving your baby up for adoption—you no longer get to care for it. It is given to the director, cast, and crew to find its way in the world. Writers must get used to not taking their work too seriously and not getting too attached to their "babies."

But despite their low placement on the Hollywood Food Chain, writers are a very necessary part of the moviemaking process. From early Westerns on, writers have been part of the Hollywood picture. Many people envy writers, as theirs looks like a pretty easy job. Hey, work a couple of weeks at a typewriter, produce a finished product of something around 120 pages, take a couple weeks off and go to the Bahamas, and return to some more deals that need to be made and begin writing another script. The reality is that only a few screenwriters are employed on a steady basis, and most free-lancers have other ways of making money, which helps them with their financial challenges during the time between writing gigs. Being a writer has been the dream of many in Hollywood history.

BABY WRITERS TO SCRIPT DOCTORS

Due to the very nature of this job, writers work on their own. In the early days of the studio system, writers were herded into office buildings where they would work at rows of desks, waiting for directions from the studio's head moguls, writing script after script for a musical or a feature being shot coincidentally on the lot. This mass gathering of writers broke up when the studio system broke up, and writers were then given their own offices on studio property. Many of them were allowed to work there year round. This arrangement changed when space became sparse, and writers were then allotted an office only when working on a specific project—and for a lim-ited amount of time. With the birth of independent films and their produc-tion, writers became stay-at-home workers, and to this day, many work off the studio lots in the comfort of their own homes or personal offices.

Writers occasionally work with writing partners. That is the extent of their collaboration with others during the process of writing a script. This creative career in Hollywood doesn't follow the same route as many of the others. In this profession, you don't have to climb the Hollywood Food Chain to achieve success (or failure, for that matter). It is a career for only the most independent, for the writer needs to be self-motivated—no one else will do that. "The writer writes," to quote the catchphrase from the film *Throw Momma from the Train*. There are only a few different levels of writer in Hollywood and they are as follows:

Screenwriter

The screenwriter writes feature films. He composes scripts of 120 pages or less, of dramatic or comedic content. Once he is known for a specific genre of film, he will be pigeonholed into that genre. Often, today's screenwriter has a manager and an agent to handle his career. If he writes a spec script (a script written on speculation, not for any particular producer or company) that sells and is a success, his career is off to a jump start. If he writes a mediocre screenplay, he will continue to write a number of scripts and have an average career in hopes of cashing in on a box-office success. Once that happens, he is gold and he won't have to worry about his career ever again.

Script Doctor

A script doctor is a screenwriter who has usually had some moderate success (and perhaps mega–box office hits, but this is not necessary) with his screenplays and is called upon by a producer or studio to rewrite a younger or less seasoned writer's script. Script doctors are paid enormous amounts of money, for it is thought that they can put their Midas touch on a script and make it 99 percent better. Script doctors are really just executive's pawns, though, brought in when a project is in trouble (i.e., written badly). It is thought that if the executive spends hundreds of thousands of dollars on a script doctor's rewrite then the project will be saved and ultimately a total success. Script doctors are exec's insurance plan. Many times script doctors receive well into the mid-six figures for the rewrite of merely one or two scenes. This is a very lucrative position to get into.

Baby Writers

Baby writers are usually young writers, or first or second-time writers who are brought in to rewrite a script when the production cannot afford a script doctor. These are serious writers, who take a stab at rewriting a script that is seemingly in trouble. This is a nice opportunity for young or not-so-mature writers to establish themselves as writers in their own right or upcoming script doctors. Baby writers are, of course, individuals of legal age—anywhere from eighteen to fifty-five—don't let the name mislead you.

Overall, writers are writers, and they can work independently or exclusively with a producer or studio. The level of burnout is high. Writers need time and space between projects in order to be fresh and productive. However, once they sell their first major script, it is to their advantage to stay in the limelight as long as possible, producing screenplay after screenplay, for it is often that a career crashes and burns once the writer has peaked with one or two major projects. The establishment of each writer's oeuvre is extremely important, as he will be branded and known for his expertise. Capitalize upon that energy and stay with it—become the expert of that genre and your writing career could last a lifetime; make a mistake and miss capturing your own brand and essence, and you will find yourself all over the place, scattered and unemployed.

SCRIBES OF THE SCREEN

In order to understand the job of a writer, let's look at some of the famous writers of the big screen. From Humphrey Bogart's Dix Steele in *In a Lonely Place* to Kevin Bacon's just-graduated filmmaker-writer Nick Chapman in *The Big Picture*, these characters will show us what the life of a working Hollywood writer is like.

Boy Meets Girl (Warner Bros., 1938)

Boy Meets Girl pokes fun at two screenwriters who repeatedly use the theme of "boy meets girl—boy loses girl—boy gets girl" in their pictures. James Cagney and Pat O'Brien are teamed as writers Robert Law (Cagney) and J. C. Benson (O'Brien). *Boy Meets Girl* was a box-office success in 1938. Only people who had worked in Hollywood would have been able to write this biting satire. It is rumored that the movie was based on the lives of

Ben Hecht and Charles MacArthur, although these famous writers have never acknowledged the fact.

The movie exposes the manic lifestyle of screenwriters working hard in the thirties. They display a sign DANGER HIGH VOLTAGE—MEN AT WORK as a warning to anyone who comes near their office. These two have been fired from every other lot due to their pranks. Benson is a man who has worked his way up from studio painter and prop boy. Law is a frustrated novelist who admits, "We're not writers, we're hacks . . . my God, I wrote once, I wrote a book—a darn good book. I was a promising novelist . . . and now I'm writing dialogue for a horse." Both are veterans at every formula and cliché in the trade. Prototypes for Jerry Lewis and his physical comedy, Cagney and O'Brien provide an accurate look at what it was like to work as screenwriters in this madcap business.

Hearts of the West (MGM/UA, 1975)

Glimpses of the mythic Hollywood that was, including the C-budget tap-dance musicals, lunch breaks of an awful costume drama, and visits with the casts and crews of the thirties B-movies are featured in this 1975 film about naïve Hollywood hopeful Lewis Tater (Jeff Bridges). This film accurately meshes reality, Western mythmaking, and movie production, and was shot on the very locations where the original story took place in the thirties. One of the featured locations is Gower Gulch, just off of Santa Monica and Gower, the center of B-Western moviemaking where hundreds of cowboys, some real, and some not-so-real, roamed the streets waiting for movie work.

Iowa-born and -bred Tater is convinced that he can go to Hollywood after he has learned to write Western pulp fiction through a correspondence course. He is a true cowboy at heart, but he is fifty years too late to be a real cowboy. Now he must settle for the movies that feature the way cowboys once were. And along with the reality of the situation, he learns the ways of the world—he grows up. After becoming involved with two con artists in Nevada, Tater is soon rescued by a low-budget movie company and befriended by Howard Pike (Andy Griffith), an extra in the movie. Tater joins Howard's crew as a stuntman and finds himself portraying a cowboy in the movies, the very thing he wants to write about. He becomes smitten with screenwriter Miss Trout (Blythe Danner). Tater then writes a novel titled *Hearts of the West* and trustingly shows it to Pike, who in turn sells the property to a studio for production, which reminds us that even in the thirties the stealing of writer's works was commonplace. Tater eventually makes it out of a bad situation, and his journey as a writer-stuntman is an important look at Hollywood of the thirties.

Here is a perfect portrayal of a man who follows his dream and not only writes his novel but becomes part of the entire story he is writing about. There are many levels at work in this film as our main character writes about his experiences and we the audience watch him at work through the very medium of moviemaking he is exposing in his novel. Tater is told he lacks nothing but confidence, but blindly he persists only to realize his Hollywood dream and end up with a published novel and the girl. *Hearts of the West* is a beautiful reconstruction of a slice of bygone Hollywood life.

Barton Fink (Twentieth Century Fox, 1991)

It is 1942, and Barton Fink (John Turturro), a critically successful New York writer, is offered a job in Los Angeles as a contracted screenwriter for Capitol Pictures. Although Fink gets involved in an eerie nightmare, the scenes of him struggling to write a simple B-movie script while staying in a seedy Hollywood hotel (the Hotel Earle where you can stay for a day or a lifetime) are unforgettable. In Room 621, for $25.50 a week, resident Fink experiences macabre events, both real and imagined. He has left his family and friends and entered the whimsical realm of Hollywood. For a $1,000 a week he is contracted to write a screenplay, something he has never done before, and soon he acquires writer's block. "What kind of scribbler are ya?" asks his neighbor Charlie Meadows (John Goodman). Fink is certainly confused. He spends hours just staring at the typewriter. He has no ideas. He begins to drink, and when a murder is committed in a nearby room, his notions of writing something with "that Barton Fink feeling!" fall by the wayside. It will be a wonder if he writes anything at all.

Fink's writing journey is an accurate portrayal of the angst a writer goes through when called upon to write a screenplay version of his stage play or novel. Often authors coming from these other media actually have no idea how to translate the words from the page to the large screen. Fink's passion is there, but it's not enough to overcome his block. The strange miasma he finds himself in jump-starts his inspiration, and at the end of the film, Fink's world returns to sanity, leaving the audience to guess what was in Charlie's box of inspiration (which some have supposed to be the head of the murdered body). Nonetheless, the Coen brothers have provided a look at a Hollywood writer and his struggles in the Hollywood of the forties.

Without Reservations (RKO, 1946)

Cut to a 1946 film featuring Claudette Colbert as author Christopher "Kit" Madden, a book-smart writer who has written *the* novel of her day, titled *Here Is Tomorrow*. Kit is just about to board a train from New York to Hollywood when she learns that Cary Grant has dropped out of the studio's production plans, and she must find someone to play the lead in her movie adaptation of her novel. (Having the novelist do the casting is an anomaly— it was rare in the forties, and it's rare now.) Enter John Wayne as the Marine Rusty Thomas, and America's Joan of Arc, as Kit is referred to, is swept away with romance and intent—the intent of having Rusty play her lead on the screen.

The train ride takes off, and Kit never reveals her true identity to Rusty and his buddy Dink (Don DeFore). Through the entire trip across the states, Kit and Rusty have discussions about life and the basic attraction between males and females. Kit learns a few things that cannot be found in books, and by the time she gets to Los Angeles, she tells Rusty the truth about her identity, losing her leading man for the film—but not for her life. For a change, this forties comedy takes a look at a female writer and has a few good points to make about gender issues along the way. Kit is a confident writer, who stands by her truths. She boards the train thinking she has the whole male-female thing figured out—until a true gentleman shares with her his thoughts about the male side of romance. With her newfound information, she is able to gain a better understanding of life and the basic attraction between males and females, becoming a softer, more feminine female writer along the journey.

The Way We Were (Columbia, 1973)

Usually not thought of primarily as a film about Hollywood, *The Way We Were* should be mentioned in this section due to the fact that the main base of the storyline involves a Hollywood writer. Here author Hubbell Gardiner (Robert Redford) moves to Hollywood after writing one good novel and is forced into writing a screen adaptation of the novel. He settles into a profitable life as a screenwriter in forties Hollywood but eventually finds himself a witness at the House on Un-American Activities Committee investigations. It is here that he and his wife, Katie Morosky (Barbra Streisand), become entrenched in different political ideals and separate only to be reunited much later, each with a new life.

The Way We Were is primarily a love story. However, its backdrop is Hollywood in the forties and it does provide a glimpse at what that life was like. Writer Hubbell has no choice really but to be part of the cookie-cutter writing that took place during this time, when studios would not take any chances—for political reasons, but mainly because Hollywood now had some very real competition: the appearance of television in every household.

SECOND-HALF-OF-THE-CENTURY SCREENWRITERS

As we continue into the fifties, the following two movies, both produced in 1950, happen to capture the pure essence of the quintessential screenwriter in Hollywood. Both Dix Steele and Joe Gillis embody all of the Hollywood screenwriters who have appeared in the movies before them and provide a prototype for those who will follow. Here, now, are two of the strongest characters the movies have ever known. Interestingly enough, they are characters who write movies.

In a Lonely Place (Columbia, 1950)

In a Lonely Place starring Humphrey Bogart is a famous film of the film noir genre. Bogart is outstanding as a Hollywood screenwriter who's got a criminal record—and a fascination with killing. He's Dixon "Dix" Steele and he makes the mistake of asking a hatcheck girl, one Mildred Atkinson, home to read to him. Yes, read (just read, nothing else) to him. He asks her to read a novel that a big director wants him to adapt into a screenplay. Mildred innocently does just that and leaves Dix's apartment. When Mildred is found murdered the next morning—in Benedict Canyon (site of the Manson murders years later), Dix is the prime suspect—he's cleared (somewhat) when Miss Laurel Gray, a confident Gloria Grahame, tells the police that she saw the girl leave Dix's apartment by herself.

One of the most interesting parts of this film is the apartments themselves—they become so involved in the story it's as if they are another character. Director Nicolas Ray, who made *Rebel Without a Cause* shortly after this film, shot Dix's world in black-and-white. The result is a dark, cold environment—a perfect place for a murderer to live. The style could be called California baroque, a Hispanic style known as "Neo–Leo Carrillo"— the scenes are counterpoint, never head-on, while the main dramatic theme is developed in space and time. This is Ray's signature style. The story keeps upping the ante against Dix until even Laurel, who has now fallen in love with Dix, questions his innocence. All the while, Dix keeps writing—sometimes all night. Laurel is his good-luck charm—he's never written so well— but all of this changes as the murder engulfs Dix's life and one of the best films noirs in history unfolds.

This film also features a catchy phrase of dialogue, repeated throughout by Dix who doesn't know where to place it in the script he is writing: "I was born when she kissed me, I died when she left me, I lived a few weeks while she loved me." It turns out that he lives and dies by this exact line by the time the movie ends.

In a Lonely Place doesn't feature the group scenes and energy of Hollywood as do some of the other films mentioned in this chapter, but it does echo the pathos of Hollywood. Dix Steele's haunting recreation of how the murder could have happened will stay with you well after you've watched this movie. It is his capacity to invoke fear that makes him the excellent screenwriter that he is.

Sunset Boulevard (Paramount, 1950)

Joe Gillis (William Holden) is a struggling screenwriter living in the Alto Lido Apartments in Hollywood. The time is the early fifties. Silents have been silent for years. Movies have become a staple for Americans. Joe is the quintessential all-American screenwriter and a classic film noir character. He's a young writer from a generic Midwestern town. He has dreams of making it in the movies. He's handsome. He's savvy. He's confident. And he's dead—at least when the audience is introduced to him and he begins to tell his story.

Joe's career is typical. A few important Hollywood suits know him. He hasn't had a writing job in a while. He's pitched every story he's got. His success has been marginal. He seeks financial help from an agent and a studio head, but to no avail. The guys from the finance company are after him to repossess his car. ("If I lose my car, it would be like having my legs cut off.") A chase ensues, and a flat tire causes him to turn into the Sunset Boulevard driveway, which alters his life forever. He hides his wheels in the garage of a spooky decaying old mansion and enters the big house where a stately older German butler ushers him upstairs—Madame is waiting.

Like so many other tragic heroes, Joe Gillis doesn't have many options when he meets Norma Desmond and her butler, Max. He learns that Norma has a script (doesn't everyone?). He tells Norma he's a screenwriter and "The last film I did was about Okies in the Dust Bowl. You'd never know it, because by the time it reached the screen the whole thing took place on a torpedo boat." Norma offers him the job of rewriting her script and although he is righteously indignant, Joe's financial situation forces him to accept her offer to move into the room above the garage and do the rewrite. The fallen star becomes delusional and falls in love with the young, handsome Joe. He tries to escape the creeping paralysis of the house and fails completely. When the TV news arrive to report on Joe's death, his body floats in the swimming pool, while the deranged Norma, about to be led away by the cops, primps for the camera ("I'm ready for my close-up, Mr. De Mille"). Tragically, Joe is the embodiment of a Hollywood writer circa 1950.

And One More from the Fifties

Susan Slept Here (RKO, 1954)

Susan Slept Here is narrated by talking Oscar statuettes, who explain to us that right around tax time every year Hollywood throws itself a surprise party, and they call it The Academy Awards. This gem of a movie oozes with squeaky-clean, all-American, good-natured fun from the fifties and features a very young Debbie Reynolds as the seventeen-year-old lead, Susan. Mark Christopher (a very wonderful Dick Powell) leads a polished but dull life as a Hollywood screenwriter who has had a few hits—and an Oscar. One Christmas holiday, juvenile delinquent Susan is dropped into his life when the cops find her and she has nowhere to go. Mark's cop friends know he's writing a story about juvenile delinquents, so voilà! she appears on his doorstep. (Believe me, this film would never be made today. Think about it, this man is in his forties, harboring a seventeen-year-old girl on Christmas Eve. That would be called something else in this post–*Leave It to Beaver* era.) At any rate, this movie was called a "cute," sexy comedy as Susan learns from her screenwriter mentor all about growing up and being a real woman. The banter between them is clever, and the acting is engaging. Worth the watch, just to hear the witty dialogue between them.

POST-MODERN AND POST–FILM SCHOOL WRITERS

For the first half of the twentieth century, writers in Hollywood generally came from other fields, such as writing novels or stage plays, and found themselves writing for the screen unexpectedly and without any previous screenwriting experience. In the second half of the filmmaking century, the films feature modern-day screenwriters, who are generally graduates of film schools, unaware of any form of writing other than for the screen.

Paris When It Sizzles (Paramount, 1964)

Ah, once again Bill Holden returns to the screen as screenwriter Richard Benson, only this time he refers to himself as a "screenwriter of stature, a famous international wit" as he holds forth in a luxurious hotel room in Paris. Enter the beautiful Gabrielle (Audrey Hepburn), the free spirit who is going to type up his 138 pages of scribble and unknowingly help him write the rest of his ill-fated plot along the way.

This delightful movie takes place during a weekend, as Benson has to have the script in two days. As he reveals the script's subject matter and the title— *The Girl Who Stole the Eiffel Tower*—Gabrielle reveals her life story. She arrived in Paris two years before and, as she exclaims, "I came here to live!" She doesn't go to bed before eight in the morning and is seeking experiences—life experiences. She's dating an actor and begins to recall how she spent Bastille Day with him. Benson likes it, he likes it a lot, *and* he is beginning to like her a lot. As the two talk about their experiences and fantasies, they find themselves acting them out, and vignette after vignette appears. They portray the elements of screenplay writing, as each scene dissolves or credits appear on the screen. Who needs Sid Field to learn how to write a screenplay—just watch these two!

Finally, Benson admits that a writer's life is a terribly lonely one. Gabrielle replies coyly with "Have you any idea what happens next?" And, of course, they fall in love, sixties-movie-kind-of-love where everything is Technicolor-perfect and they live happily ever after. End of story. Amen.

The Big Picture (Columbia, 1989)

The Big Picture is a good-natured satire about Hollywood that explores contemporary mores through Sir Christopher Guest's keen eye for the silliness of the film business.

This dead-on spoof of Hollywood premiered at UCLA, and its newspaper ad showed Kevin Bacon in a shopping cart with a camera in hand and the words FILM SCHOOL PREPARED NICK FOR EVERYTHING . . . EVERYTHING BUT HOLLYWOOD. This is not the dark side of the business, like *Star 80*, or even the subtle sarcasm of *The Player*. This is a very real scenario of what a film-school grad will face upon arrival in Tinseltown.

That new breed of filmmaker is portrayed as Nick Chapman (Kevin Bacon), a writer-director fresh out of film school, wins the NFI (National Film Institute, an American Film Institute lookalike) trophy for his student film *First Date*. Everyone who is anyone in Hollywood is after him. No one has seen his work yet, but he is told that he is "brilliant," "marvelous," and "sensational." He has been successful at creating a buzz about himself. A surprisingly real display of making it as a writer-director is depicted on screen, as Nick's journey to see his concept become alive on the big screen is unraveled.

Nick meets with studio executive Alan Habel (J. T. Walsh). "Tell me your movie, Nick." Nick begins to relay his storyline. His story is then played out in black-and-white. When Habel interrupts Nick, asking him to change the love-story triangle of a woman and two men to a woman and another woman instead of a man, Nick is hesitant. His pitch is okay, by most standards, and now he's been thrown for a loop. He'll think about it—the change, that is. And so he does. As Nick becomes engrossed in the trappings of the glamour of Hollywood—i.e., starlets who want to be in his movie, offers from other studios, story meetings, new cars, a new apartment, rewrites—his relationship with his girlfriend and old friends suffer. When Habel gets the hook and is let go at the studio, Nick learns the real truth about working in Hollywood—no one knows you when you're down. He loses his deal and, along with that, all of his luxury items. He is forced to work as a messenger. Eventually, he ends up directing a music video for a friend. Habel's office sees the video, likes it, and once again calls on him to do a movie for them at their new studio. Nick agrees to work with them but only if he can do his movie the way he wants to do it.

Nick has learned his lesson. If anything, this film gives hope to all those who pursue a career in Hollywood, for it outlines the reality of having your vision trampled on by egocentric producers, insane agents, and all of the "circus performer" types that populate Hollywood, but it also shows a character ultimately winning by exercising his independence and his integrity.

The Player (Fine Line, 1992)

This film has played a part in a number of different chapters in this book. In *The Player*, it is the writer who poses a series of questions and problems to main character studio head Griffin Mill (Tim Robbins). Writer Michael Tolkin himself is an example of that rare writer who has been successful at the writing of both novels and screenplays. He is one of only a handful of folks that include John Irving, Amy Tan, and John Grisham, who are part of the all-types-of-writing-club. In this film, Tolkin chooses the writer character, David Kahane (Vincent D'Onofrio), to be the one who makes a statement about the abuse writers take from other players in the industry. Through the Kahane character, Tolkin lashes back at the industry. Being a writer, Tolkin uses Kahane and his actions to make a statement about what he'd like to do to development and studio execs. *The Player* entertains on many levels and it provides a look at what happens when the Hollywood writer becomes a villain. In the nineties, the writer in Hollywood has come to this, a far cry from those pranksters in *Boy Meets Girl* or *Barton Fink*.

My Life's in Turnaround (Islet, 1994)

Two New York City out-of-work writer-theater junkies—roommates Splick (writer Eric Shaeffer), a cab driver, and Jason (Donal Lardner Ward), a bartender who fantasizes about underage models—find their lives going nowhere. One morning, Jason goes to Splick's room and wakes him up. He proceeds to ramble off the names of some of the more famous independent filmmakers in history, and from there they decide that they are going to be filmmakers—never mind that they don't really like films. They seek out a friend who works at a junior talent agent in hopes that she'll set them up with meetings and they'll proceed from there to be writers, producers, directors—whatever it takes to get their movie made.

This film is mentioned here to show that one does not need talent to write a screenplay and get a movie made. One only needs drive and ambition and to essentially just "show up"—just like these two.

Get Bruce! (Miramax, 1997)

In many ways, this documentary portrait of Muppet-like comedy writer Bruce Vilanch, a Hollywood staple, a veteran writer of most of the major awards shows on television, is an upbeat look at one writer's life, albeit a gay writer's life. Vilanch is a colorful character, and this doc shows him at home, interacting with his mother, and working with all of his famous friends, such as Bette Midler, Billy Crystal, Whoopi Goldberg, and Robin Williams, to name a few. Vilanch's creative process is explored, and it is purely a slice of life only the documentary film camera could capture. It is the most accurate of movie depictions of Hollywood writers. However, it also has such high energy that it often seems like a fictional film—no one could be this outrageous in real life.

Honorable Mentions

Best Friends (Warner Bros., 1982)

The Lonely Lady (Universal, 1983)

The Muse (October Films, 1999)

Look to the eighties for two very mediocre films about writers, *Best Friends*, starring Goldie Hawn and Burt Reynolds as married screenwriters, and *The Lonely Lady*, which stars Pia Zadora as a struggling screenwriter. Both of these films expose the usual ups and downs of the screenwriting life, making it seem very ordinary, actually. And finally, the last entry of writers' movies of the twentieth century is Albert Brooks's *The Muse*, which is mostly a culmination of every Hollywood cliché seen in every other movie about Hollywood writers. Nothing new here, and certainly nothing that illuminates or helps one to understand the profession.

IN REAL LIFE

An energy shift has occurred during the last century. Writers are now fed and taught by the media, which includes print, audio, and visual stimulation.

Self-Segregation

One element that remains constant through the century of filmed writers' lives is their need to be independent. Of the movies featured in this chapter, only three films, *Boy Meets Girl, Best Friends,* and *My Life's in Turnaround,* focus on a team of writers. Benson and Law are a couple of guys in the thirties who have been kicked out of every studio—pranksters, for the most part. Today they would be working in the genre of comedy features, or at the very least be gainfully employed as television sitcom writers. And those two in New York work as a couple of Gen-X slackers, and the husband-wife team is credible. The remaining films feature very memorable lead characters who struggle with their art, the art of writing.

Author Otto Friedrich, in his book *City of Nets,* an excellent look at Hollywood in the forties, states: "Hollywood really is an imaginary city that exists in the mind of anyone who has, in his mind, lived there. My Hollywood is different from your Hollywood, just as it is different from Rex Reed's Hollywood, not because they know more about Hollywood than you or I do but because they are different from us, just as we are different from each other. No matter when one lives in Hollywood, one brings one's own mental furniture along."[2] In addition to the mental furniture, these writers had one foot in reality and another in fantasy as they not only made up their storylines, but their own lives as well. Often, the journey taken while they were writing led them to new stories, stories about writers writing for the movies.

This progression of movies through the decades shows us how much the industry has changed, and how those changes have affected the position of the writers. Back in the golden days, writers generally only talked to writers who worked for the same company—otherwise, their idea could be stolen. A $500-a-week writer would not be welcomed at a $1,500-a-week writer's party. Self-segregation was born. It wasn't safe to share ideas, for they could be stolen within minutes. This is a practice that continues to this day, and this is why High Concept was born. A High Concept—the combination of two or three well-known and successful

[2] Friedrich, Otto, City of Nets. Harper & Row, New York, 1986, p. xii.

movies in one quick catchall sentence—to suggest the essence of your script is used to talk about your story. This way, you are not giving it away to everyone who hears it.

In a 1996 article in magazine *Entertainment Weekly,* the question was asked, "Who killed the Hollywood screenplay?" An answer was immediately suggested by the following paragraph: "The writers of the thirties, forties, and fifties were frustrated novelists. They grew up in an atmosphere of storytelling. Today we live in an illiterate culture. The screenwriters today get their education from television. Nobody reads books anymore,"[3] and we have seen moments when our writer-heroes have questioned their hand at writing screenplays for exactly the reason stated here—that perhaps they were too literate?

The Totally Visual Writer of the Screen

As the seventies and eighties progressed, film schools began to churn out novice screenwriters—screenwriters who were raised on television, not all that familiar with literature classics and the history of English composition. A new breed of movie writer was born. So, by the time we meet Nick Chapman in *The Big Picture* we should know that this writer may or may not be trained as a novelist or fiction writer. The art of screenwriting makes a shift in the second half of the century, as seen in the stories of these Hollywood writers.

As we move into the twenty-first century, writers will be versed more and more in the visual culture and influenced by MTV-style media, and less and less familiar with the old-school design and discipline of the written word and books.

TWO YOUNG SCRIBES

Here are two young writers' voices.

---

[3] Svetkey, Benjamin "Who Killed the Hollywood Screenplay?" Entertainment Weekly, October 4, 1996.

Two Young Scribes

J. Ryan, Writer

J. Ryan eats, drinks, lives, and sleeps the writer's life. He has the look of the early Beat Writers, a quintessential squint of questioning remains ever on his face as if he is in constant thought—thought about how he can make his everyday life and all that appears in front of him work in a scene in one of his latest screenplays. He is twenty-five years old and has written numerous screenplays and stage plays to date. He is dedicated to being a full-fledged screenwriter and is in search of being discovered.

Have any movies about working in the industry had any influence on you?

I certainly wasn't swayed to work in entertainment by movies like *The Big Picture* or *The Player*. Despite the uniformly negative portrayal of screenwriters in Hollywood-themed films, the whole process of filmmaking, even from a distance, seemed vain and lugubrious. It wasn't until I moved out here (L.A.) and actually got involved in it that I found that it's a great deal of fun as well.

When I was a teenager, I started to believe that the storytelling potential of cinema was unmatched, and I arrived at the opinion after seeing *Midnight Cowboy, The Godfather*, and, of course, *Star Wars*. These are the films that probably influenced my decision to major in radio, TV, and film [in college] and concentrate on screenwriting. Then I got to college and started watching movies like *Who's Afraid of Virginia Woolf?* and *Streetcar Named Desire*, and discovered theater, which totally blew my naïve, corn-fed Midwestern brain.

I love a good story, I'm a loquacious person, and generally pointless and postulating in conversation, so I've always had a deep respect for those who can focus their own wild imaginations enough to tell a compelling tale. My favorite movie of all time is still *Midnight Cowboy*, because of its skill and grace with a gamut of emotions. In recent cinema, I always think about the opening sequence of *Raising Arizona*, the middle third of *Crouching Tiger, Hidden Dragon*, and the bulk of *Amores Perros*. And I wouldn't be where I am today without *Pee-Wee's Big Adventure* and *Back to the Future*, not by a long shot. Thank you, movies.

Do you see yourself as someone who can predict what audiences will want?

No way, that's a fool's errand. By the time you finish your screenplay in today's "hot" genre, something else has replaced it. I'd rather try to make something good, which is also a fool's errand, but I sleep better for it.

Do you write for yourself or for your audience?

I'd be utterly flattered if someone besides me were even vaguely interested in what I have to say. And also impressed that they were able to make sense of it. I shouldn't write for other people; it's pejorative.

How about a combination of writing for yourself and your audience?

Ask me again in a couple of years.

How big a part does competition play in your writing—are you concerned that others may come up with the same ideas you have—or steal them?

Ha! I've never heard of anyone wanting to steal any of my ideas. If someone really wants to, I say go ahead, I'll make more. You'd still have to go about the near-miraculous task of getting a film made, and I don't envy you that.

What's been the toughest thing about being a writer so far?

Selling yourself. I've never liked that. I always want to let my writing do the talking. I really have to get over that though.

Martin Fletcher, Screenwriter

Martin has had success as a screenwriter working with a major comedian-star to write and punch up projects the comedian-star has in development at a number of studios and networks across town. He's learned firsthand about the business through pitch meetings, agent get-togethers, and social gatherings with his star connection.

Are there any films that influenced your decision to be a screenwriter?

My favorite movie is *The Godfather.* I always wanted to be Michael Corleone; however, my family is neither Italian nor in the mob, so I had to just fake it. L.A.'s the only place where that type of behavior is not only acceptable but encouraged.

Do you write for audiences or for yourself?

Well—I think that oftentimes predicting what audiences will like is the job of the studio executives. I've had the chance to work with some execs (as a lowly assistant, not as a writer), and on several occasions, I've questioned decisions that they've made and then watched as the movies they were making, while not very good, still made piles of money. I'm not sure that that's a skill that I have. I usually write for a combination of myself and the intended audience. There is a definite set of writing rules, especially in film and TV, that you must learn and understand and incorporate in your writing. The film audience is rather sophisticated and enters every film with certain expectations based on the genre of film they are seeing. Our first job is not to cheat the audience. If they have expectations, and you do not meet those expectations, they will be disappointed. The best writers learn to break those rules and keep the audience happy at the same time.

Does competition play a part in your writing? Are you concerned someone may steal your ideas?

I don't worry about that too much. My ideas aren't very good, not really worth stealing.

What's been the toughest thing about being a writer so far? The writing itself or selling yourself?

I think that it's the writing. It is not easy to write a great screenplay. There are so many elements involved that must really come together. Through the magic of Hollywood, there are a lot of bad writers who somehow break through and make a great living, but I don't think that true talent goes unrecognized. I feel like oftentimes people who complain about not having access just don't have a great product to sell. They may have something better than what they see in movies or on television, but that certainly doesn't mean that they are great writers. On the other hand, I've been writing with my writing partner for over a year for an actor that now has his own TV show, and because of politics and other issues, we can't get staff jobs even though we feel we are fully qualified. That type of thing is frustrating. So, really, both aspects have been fairly difficult.

CHAPTER 11

LOT LIFE

It is neither a factory nor a business establishment nor yet a company town.
Rather it is more in the nature of a community, a beehive, or,
as Otis Ferguson said, "fairy-land on a production line."
—Carey McWilliams, Southern California: An Island on the Land

Studios are like little cities onto themselves. Most of them have their own zip codes, general stores, water supplies, security departments, and commissaries, to name just a few amenities. They are generally walled in for protection against any intruders and are known to have unique histories. They have entrance gates and back lots, reputations and certain standings in the community. The eight major studios employ many Los Angeles–area residents. They are (in no particular order) Viacom/Paramount, Sony/Columbia, Fox, Warner Bros., Disney, MGM, Universal, and Dreamworks. All of the above have actual studio lots on property within the Los Angeles area, except for Dreamworks, which shares space on the Universal lot.

A BRIEF HISTORY

Previously, in the chapter dedicated to production and crew personnel, we discussed the journey many individuals made to the Southern California area to work in this new motion-picture industry. A brief, thumbnail sketch of the studios goes like this: By the end of World War I, three studios flourished. They were the Lasky Corporation, Paramount, and Lowe's International. By the twenties, movies, distributed throughout the world, became the chief avatars of American culture. During the twenties, the three original studios metamorphosed into eight new ones. They were First National, United Artists, MGM, Twentieth Century Fox, the PDC (Producers Distribution Corporation), the FBO (Film Booking Office), Laemmle's Universal, and Warner Bros. With the introduction of sound in 1927, MGM, Paramount, Warner Bros., Fox, and RKO began to emerge as the five major studios, leaving in the minor leagues Universal, Columbia, and United Artists.

By 1933, hard-hitting effects of the Great Depression devastated the industry. After that financial breakdown, MGM soon ranked number one, mostly due to the hard work of Irving Thalberg. Paramount became the most "American" of the American studios but soon started employing European directors, DeMille being one of them. Warner Bros. became known as the studio that represented the working class, and Universal was famous for its horror movies. Columbia's success was attributed to Henry Cohn, and UA and RKO distributed independent films known as B-movies. All other small studios and production companies were considered B-moviemakers, famous for distributing their movies under a double bill so the audience would stay for the second film even if it were as bad as the first.

THE DREAM FACTORIES

Many of the original studios are currently completely functional and located in the area surrounding Chaplin's original studios. They are:

- **RKO Studios**—at Melrose and Gower
- **Raleigh Studios**—at 650 North Bronson
- **Hollywood Center Studios**—at 1400 North Las Palmas
- **Paramount Studios**—at 5500 Melrose (look through the grand entrance gates on Melrose to see a direct view of the Hollywood sign)

- **Warner-Hollywood Studios** — at Formosa and Santa Monica
- **ABC Television Center** — at Prospect and Talmadge

This chapter celebrates these special parcels of land where so many movie memories have been produced. A studio is a metropolis in its own right, a place on earth that yields to dream-making. The movies that have incorporated the studio into their storylines will provide us with an inside look at these magical places. Many of the following films have made the studio itself a main character in the film.

As the main characters in the movies discussed in this chapter have their studio adventures, many lot workers will make appearances in the scenes. We'll be taken on these tours by journeymen of the studio, workers who have realized their dream of working in the movies.

Free & Easy (MGM, 1930)

Buster Keaton appears in his first talkie as Elmer Butts, a young man from Kansas who travels to Hollywood with Elvira Plunkeet (Anita Page) newly named "Miss Gopher City." Upon arriving in Hollywood, to the MGM lot, Elmer meets up with his friend Larry (Robert Montgomery). Larry has achieved leading-man status. He likes what he sees when he meets Elvira, so Larry and Elvira sneak away to be alone. This leads Elmer on a joyful jaunt through the movie studio with the hopes of getting Elvira back, and a cook's tour of MGM ensues.

As talking films emerged, the comic masters of the twenties were replaced. Keaton, Chaplin, Lloyd, and Langdon made room for the antics of the Marx Brothers and the verbal con men of vaudeville. With this film, however, Keaton becomes "human," as audiences hear his voice for the first time (it resembles that of the comedian Tim Allen). He disrupts the production of several films in the making, negotiates soundstages, and encounters various celebrities, such as Lionel Barrymore, Jackie Coogan, and Cecil B. DeMille.

Unfortunately, Elmer does not get Elvira but he does land a contract to star in screen comedies. So, the movie ends happily professionally, not romantically, but then, that's a pattern we've seen in earlier chapters and will see again here.

Movie Crazy (Paramount, 1932)

Turning up the volume just a bit, *Movie Crazy* has another young man from Kansas interested in working in the movies. This time, comedian Harold Lloyd is Harold Hall, and he has high hopes of being a star, but upon arriving at the Hollywood studio, he is only able to find work as an extra. From the moment he steps off the train and onto a production within the lot, he manages to get in the way of everyone he comes in contact with.

Aggravating the director, he takes a shine to the leading lady, played by Constance Cummings. She is inviting at first but later turns him down by writing a farewell note on the back of a party invitation. Harold thinks he's been invited to the social gathering and makes an appearance. While in the men's room, he accidentally dons a magician's jacket and returns to the dance floor. One by one, doves and rabbits and magic oddities emerge from his coat as he dances. This scene is movie magic at its finest, a bridge between Lloyd's physical silent comedy and this new brand of audio movie. Harold continues with his usual antics, and he, too, loses the girl but is awarded a studio contract to star in future movies.

Singin' in the Rain (MGM/UA, 1952)

Singin' in the Rain is one of the best-loved musicals of all time and is often referred to as the apex of the American studio musical. It is included in this chapter because it is a nice scenario of moviemaking in the late twenties and early thirties, when studios were exercising their heavy hands and were faced with the futuristic monster known as sound.

"Every studio is jumping on the bandwagon and theaters are installing equipment. We don't want to be left out of it!" one of the studio execs exclaims. And while the studio top brass scramble around thinking of ways to outdo their competition, the problems the stars face are even greater.

As the spokesperson for the silent age, Norma Desmond made her famous remark about having faces; well, in this movie, the voices attached to those faces are the source of conflict. Leading stars Don Lockwood (Gene Kelly) and Cosmo Brown (Donald O'Connor) moved up the ranks of entertainment from vaudeville to stuntmen to musicians and finally to leading roles. When 1927 rolls around, and the famous first talkie, *The Jazz Singer*, appears on the screen, these guys need to reassess their talents, along with those of their female costar, one Lina Lamont (Jean Hagen). It seems Lina has the worst voice in the world . . . but the problem is seemingly solved when Kathy Selden (Debbie Reynolds) appears on the scene and her voice is golden.

Watch the conflict as it unfolds between the musical numbers. This film is famous for many things, mostly for being very entertaining, but it is also the only film that explores the advent of sound in Hollywood. Observe how the studio handles and incorporates a new technology, something that continues to this day as many new Internet and satellite applications continue to appear on the horizon.

It's a Great Feeling (Warner Bros., 1949)

Actors-directors Dennis Morgan and Jack Carson set up shop on the Warner Bros. lot to produce their own pictures. While Morgan and Carson develop their motion picture, the audience is able to see the Warner Bros. lot in action—so much so that in this case the lot does in fact become one of the characters. The movie is a comedy and features cameos from many stars of its day. The year is 1949, one year away from the film that changes the energy of the movies about movies—*Sunset Boulevard.* Consider this one the last movie that kept up that shiny, happy, false front about how wonderful it was to work on the lot. *It's a Great Feeling* lives up to its name and vigorously delivers the innocence of its time.

Who Framed Roger Rabbit (Amblin/Touchstone, 1988)

Hollywood, 1947. The golden years of making movies within the studio system are about to end. This delightful "'toon" is the only semi-animated entry that concerns itself with working in the industry; still, it has all the elements—greed, corruption, star power, murder, sex, and blackmail—to make it a legitimate movie-about-the-movies entry. The story is complex and is centered on a private detective Eddie Valiant (Bob Hoskins). Valiant has been hired by R. K. Maroon of Maroon Studios to investigate rumors surrounding his top talent—loveable box-office draw Roger Rabbit (voiced by Charles Fleischer) and his sexy wife, Jessica (voiced by Kathleen Turner). Jessica has allegedly been unfaithful, and Maroon is worried about Roger's emotional state. Valiant takes pictures of Jessica playing patty cake with Marvin Acme, the owner of Toontown, where the 'toons live. Acme is later found murdered. Roger is the chief suspect. Together, Roger and Valiant investigate Acme's case and learn that he was involved in a corrupt plan with a local Judge Doom (Christopher Lloyd), who wants to buy out Maroon Pictures, a plan that leads to Maroon's death. Ultimately, Doom's scheme is foiled and all ends well for the 'toons, including Roger and Jessica. All live happily ever after.

In this film, Maroon is the ultimate in archetypal studio executive. He looks after his talent and studio family and is obsessed with the bottom line. His studio lot is a typical one—a bustling place filled with stars, crew, execs, and directors, all working to produce Maroon cartoons in the never-ending quest for the lucrative opening weekend. The movie cartoon *was* children's entertainment in the era before television and video. The films had to be better than any other filmed cartoons around town. The two worlds, animated and live action, take on this quest to produce the finest in children's programming. Overall, *Who Framed Roger Rabbit* may be a "'toon," but it's a good "'toon" and one that features the studio lot and its varied functions.

Falcon in Hollywood (RKO, 1944)

The Falcon franchise was a well-known series of films during the forties featuring Tom Conway as the supersleuth detective Tom Lawrence, known as The Falcon. In this particular entry, The Falcon, who is vacationing in Southern California, finds himself in the company of two beautiful film stars who lead him to Sunset Studios (the RKO Studios) where he stumbles upon the corpse of the studio designer. The Falcon takes it upon himself to solve the crime. His adventures take him through every nook and cranny of the RKO lot and back lot, and some amusing interaction with the casts and crews that are currently shooting there. The scenes blend together (reminiscent of *The Stunt Man*). Audiences are challenged to identify which scenes are part of the movie being shot within the movie, and which are part of the movie they are watching.

Four Girls in Town (Universal, 1956)

Opening with a spectacular panoramic shot of Manning National Studios (the Universal lot), this 1956 flick focuses on four young wannabe actresses as they flit about the studio lot in search of love, fame, glamour, and their next favorite outfit. This movie captures the essence of California studio lifestyle. Not only do these four transplants discover the studio, but they slowly but surely let their hair down (quite literally—from beehive to flip) and become California Girls complete with Barbie-like outfits and grand love affairs to follow. All four are rejected when they fail to visit the casting couch. "It isn't a script, a script would have made sense!" one of them exclaims. A narrator picks up the story at this point, intoning "Where movie scripts left off and life begins . . . ," and we watch as all four give up the dream of stardom and return to "real" life.

The Errand Boy (Paramount, 1961)

Despite what many say about Jerry Lewis (and it seems the world can be divided between those who love Jerry Lewis and those who hate Jerry Lewis), he is an entertainment icon and a filmmaking genius. His love of the art of making movies led him to produce and shoot independent films—in addition to the international blockbusters he had been appearing in for years, which were hugely popular the world over. His home life was documented via short films about his family, or a gathering of friends in the backyard. He would shoot the film, edit it down, and hold screenings for his friends and family.

He was one of the first stars in the history of Hollywood to be given a producing deal with a major studio, Paramount, in 1960–61. During this time, the same time that *The Errand Boy*, one of the purest and best movies about making movies was made, Lewis practically owned Paramount Studios. Any film he wanted to do he had a pretty good chance of doing. He had proven in his early comedy movies—with and without Dean Martin— that he was an audience magnet.

Lewis stars in, writes, and directs this homage to everyone who has ever wanted to be part of the film industry. It is his work as a director that gained him notoriety and respect among filmmaking peers. His oeuvre of films reflects a full life as an entertainer, writer, director, and producer. Lewis is the author of *The Total Film-maker,* and he taught classes at USC in Los Angeles in the early seventies. This man lived and breathed filmmaking, and *The Errand Boy* is a snapshot of that life.

"Hollywood is the land of the real and unreal," the narrator tells us as the camera pans over beautiful black-and-white L.A. circa 1960. Within the first five minutes, the narrator illustrates a number of different movie genres the audience might be interested in — Westerns, cheesecake, suspense, brutality, or love story — providing a vignette of each one. The narrator states, "These are examples of the unreal." The camera revisits each vignette and either pulls back or turns up the sound to discover the tricks the director and cameras play to produce what looks like the "real." And so, the land of the motion-picture czars and czarists (narrator's word) will be exposed and, to prevent the movie from being a dry and stilted documentary, the narrator continues to tell us, the talents of one of the most prominent and highly intelligent idiots available will be utilized. And that idiot is, of course, Jerry himself, shown while he is having difficulty putting up a billboard on the lot, a billboard that displays Jerry's name as a credit.

The plot begins in typical Lewis-movie fashion when the studio decides it needs an undercover spy who will secretly observe everyday operations. This individual must be unknown to everyone and cannot care about the amount of money he makes. Just as one of the board members says, "There couldn't be anybody that stupid," there is Jerry, on cue, in the board room, as Morty S. Tashman, idiot. Within hours, Morty becomes an errand boy, a gopher, a production assistant on the lot being yelled at by his mail-room dispatch supervisor: "Listen and listen loud!" Morty does his job, and as he delivers and assists throughout the Paramount/Paramutual lot, he manages to disrupt the script department, personnel department, commissary, and the secretarial department — and that's just the beginning. Each of the respective divisions of the studio are hard at work with typists typing, commissary workers working, mail and delivery boys doing just that—delivering mail.

Morty continues with his job, knowing that he is observing for some reason, reporting back to his supervisors, yet he's doing more damage than good and he knows it. At lunchtime, his habit has been to go into the wardrobe department to rest. During one of these moments of rest, he is befriended by a little clown puppet. The little puppet offers Morty a lollipop, which he takes and eats as the clown falls asleep. It's a tender and sweet moment. When Morty visits the same shelf in the wardrobe department some time later, he looks for the little clown only to find Magnolia, another puppet, only this time she's a swanlike character who talks. The exchange between the misguided guy and the caring puppet is sometimes corny, but it certainly reveals the passion and frustrations that exist within Lewis's character as well as within the many millions who have chosen to pursue creative careers in Hollywood. The dialogue between Morty and Magnolia is as follows:

MAGNOLIA: My name's Magnolia. I'm from the Deep South.

MORTY: I'm Morty S. Tashman and I'm from New Jersey.

MAGNOLIA: Why are you so far away from home?

MORTY: Oh, well, I guess for as long as I can remember I always wanted to go to Hollywood and see the movie stars and all the people that make the pictures and how they make them. And I guess it wasn't uncommon that, like me, a lot of other guys my age like the movies. So, I saved up some money and one day I got on a bus and here I was, in Hollywood. And when I got here, I realized I wasn't any closer to it then I was in New Jersey. And as you know, when you're far away from something and you can't get to it, that's not quite half as bad as when you're close to it and you can't get to it, right?

MAGNOLIA (shaking her head): Yeah, that's right, I know . . .

MORTY: So, I guess I was just a little overanxious . . . so delighted and happy about working in a studio I promised them I would give them information they wanted and I can't do it . . . I even flunked spy. From the first day that I got here, I've done nothing but cause everybody trouble. I didn't mean to. I'm a gopher, I go for this, I go for that . . . there's no excuse.

At that point, Morty tells Magnolia that he has taken up enough of her time. He's tired of his own little pity party until he realizes something. He realizes he hasn't lost his marbles, he may not be that smart, but he knows puppets can't talk. Magnolia reminds him of what it was like when he was a little boy and his parents took him to the puppet show—weren't the puppets almost lifelike then? She assures him that it isn't any different now. She tells him that he enjoyed his visit with the little clown and took him at face value. He liked what he saw and believed what he liked. Their conversation isn't any different, not one bit.

Magnolia and the little clown are Morty's mentors. Thanks to their encounters, a light bulb clicks on, his attitude shifts, and the energy changes. The next thing he knows is that he's been called to star in Paramutual's next movie. It turns out that at a screening on the lot, a casting agent–executive spotted Morty in one of the scenes he accidentally popped into during one of his misadventures. The top exec liked his energy because he could communicate with the audience and he was funny. Morty S. Tashman, former errand boy, former idiot from Jersey, Paramutual's newest superstar.

What film segment better sums up the total dedication to and final realization of your dreams than this one? Lewis incorporated the angst people feel when they are first starting out in the business, when they are still wet behind the ears, greener than green, trying to make it in Hollywood. The message is that the belief you had as a child, the belief you have always had, will make it happen for you. It isn't any different now. No difference whatsoever.

PROFESSOR LEWIS SAYS

Lewis continued to deliver this theme through his art and in his book, in his classroom, and throughout his life. He stressed the need to make film. Do something. Shoot a movie, if that is what you want to do. The quickest way to find out your capacity for being a filmmaker is to determine whether or not you have something to say on film. If the answer is "no," then stop right now. He stresses the need to have a point of view. The need to have passion and determination. That film will last for a very long time. Therefore, each one of us has to find our own *Reservoir Dogs*, *Clerks*, or *Swingers*, and make that project a reality. Do what in your heart you feel is right. Make your mark. Lewis tells the following story about a film he presented to the bigwigs at the studio and how he handled their criticism.

> Then, at the preview, the studio executives began to tell me what was wrong. They turn into experts at previews. They were in a part of the theater where they couldn't hear all the laughs. They concluded I had a bomb and buried me like crazy with all kinds of suggestions.
>
> I listened carefully and made notes like a good producer. Then I took the picture back into the cutting room. I let them think we were slaving for a day and a half. Actually, we never opened a can for deletions. We previewed again three nights later. They smiled, "Now, Jer, you've got a picture." We hadn't made a cut. We had made a slight addition.[4]

This is what every individual with a creative career in Hollywood needs to understand. Believe in your work. Believe in yourself. Passion and determination will get you where you need to be. Ninety-five percent of this business is just showing up, being there, enthusiastic and ready to work. Follow that belief you had as a child and never give up.

[4] *Lewis, Jerry, The Total Film-maker, Random House, New York, 1971, p. 32.*

CHAPTER 12

WRAP-UP

And may I say a word to this new generation [of filmmakers]. Don't follow trends. Start trends. Don't compromise. Believe in yourself. Because only the valiant can create. Only the daring should make films. And only the morally courageous are worthy of speaking to their fellow man for two hours and in the dark.

—Frank Capra, upon receiving his AFI Lifetime Achievement Award

LESSONS SUGGESTED BY MOVIE-MADE EXPERIENCE

The previous chapters *Creative Careers in Hollywood* have focused on nine major professions within the entertainment industry, and a general study of crew and production jobs on location and studio lots. We have looked at more than a hundred films about working in show business; the movies themselves have illustrated what each job entails. The stories surrounding the jobs of the leading characters have captured our imaginations and hearts, but after all is said and done, you, the reader, must decide for yourself whether or not you want to be a part of the industry, to be both *in* and *of* the movies.

Surely your experiences as professionals in the entertainment industry will not be identical to the stories we have explored; however, the study of this material can begin to prepare you for your days ahead—in the office, on the lot, or in front of or behind the camera.

SHARED CLASSROOM WISDOM
EQUALS GAINFUL EMPLOYMENT

Sharing my knowledge of the entertainment industry is as important as working within the entertainment industry. In the mid-nineties, I was given the opportunity to teach at a number of fine universities in the Midwest. I didn't use textbooks, only real-world examples of experiences, documents, and projects that I had been involved with as a d-girl, writer, producer, and executive. My classes ranged from basic screenplay writing to an intense film-as-business course that attracted over fifty students per semester. I found that many students had no clue what went on during the day-to-day operations of a studio or production company. Most of the time I would end up bringing in clips from the movies that have been discussed in this book to explain how pitch meetings went for writers, or how development discussions got underway, or how producers made their deals. All of the duties of different professionals in the industry and more can be found within these scenes. And so, the idea for this book was born.

All of the students who graduated from classes that incorporated the contents of this book during the last three years of the twentieth century found gainful employment within weeks, if not days, of arriving in their chosen media metropolises.

There has always been talk of reaching VP status by the age of twenty-five, and that goal is not considered lofty in this industry. Perhaps reaching a middle-management position by twenty-five is a bit more realistic. Do bear in mind that within two years, sometimes a year, many of the students of film who studied these movies about the movies, reached that goal. At the time of this book's printing, all of these students had viable positions in the entertainment industry, both on the East and West Coasts. Some of the jobs they have already got are: VP at one of the major studios, editor for a popular cable show, junior agent at one of the top three agencies, a writer with a three-picture deal at Disney, founder of an industry Web site, and associate producer of an indie film that has been generating some buzz. Not bad for less than five years out of college. With the one-hundred-plus-channel universe, along with the introduction of Internet and broadband technologies, there are plenty of jobs to be had. Twenty years ago, when your humble author graduated from college, there were a total of three (count 'em, three) networks and only six running studios to work at. Today, that number has skyrocketed. If you can't find an entry-level job in the entertainment industry it is not because there aren't plenty to be had—it is because you are not looking hard enough.

COLLECTIVE CONSCIOUSNESS CONTINUED

As Hollywood continues to perpetuate our contemporary myths, you too will be adding to that collective consciousness by choosing your creative career. You'll be following your personal myth to do this type of work. In Hollywood, a frontier mentality still persists. Aggression underlies most people's behavior, and because of the huge, unavoidable collaborative effort, many mask their true feelings by addressing each other as "dear" or "sweetheart." A business civilization takes over, and property (the movies themselves) becomes far more important than man, and human values have to struggle hard to survive at all. Don't fall into the trap of negativity. Consider it a privilege to be part of this select group.

HOLLYWOOD: STARS AND STARLETS, TYCOONS AND FLESH-PEDDLERS, MOVIEMAKERS AND MONEYMAKERS, FRAUDS AND GENIUSES, HOPEFULS AND HAS-BEENS, GREAT LOVERS AND SEX SYMBOLS

The above is the name of a book about working in Hollywood by author-screenwriter Garson Kanin, published in 1967. In this brief description, he pretty much captures the energy of the working world of the entertainment industry. Unfortunately, the dream industry sometimes falls into a netherland of questionable activity. We've seen plenty of ups and downs, triumphs and tragedies, unrealized dreams and incredible success stories—it's all part of the industry makeup. We are lucky to have access to the celluloid scenarios that immortalize these triumphs and tragedies, for watching these journeys can help us with our own.

Throughout your professional life, one thing will remain: Art and aesthetic goals are less important than business and financial goals (this is why it's called show *business*). Mechanized creativity took over sometime in mid-century. The studio system folded, new technologies were introduced to the public, and the film-school generation of the seventies broke the pattern to have the industry evolve into the global power that it is today.

But that aside, you still need to follow your dreams and hold on to the seed of creativity that drives your desire to be a part of this creative industry. There are two schools of thought in Hollywood. One is "Don't tell anyone anything—make them work like we did," and the other is to share, encourage, and nurture the next generation of media professionals. Those in the first category are working in fear. They are afraid of ageism and actually do not have enough confidence and self-esteem to believe in their own work.

Those who are working out of the joy of sharing, work from love. Those who share will be rewarded. It is true, there is ageism in this industry, but why not counteract it by encouraging and nurturing younger workers and executives, for if they climb higher on the ladder of success than you did, perhaps, as they are going up, they will take you along, because you helped them at some time earlier in their career. Make that overall choice right now, the choice that you will work from love, not from fear.

FROM EXTRA GIRLS TO ERRAND BOYS

Historian, writer, and myth expert Joseph Campbell said, "Follow your bliss." And this you must do.

Hopefully, you are one of the few who has been ruined by the movies, just as Holden Caufield described himself in the quote in the beginning of this book. Ruined enough that you love them so much that you want to make them your life's work. In that case, dream big and use this book to guide you through the adventures of those who chose the same career before you. Learn from their mistakes and from their triumphs. They are there to inspire you to do a better job, live a better life, and produce entertainment that reflects the public's collective consciousness while resonating in the public's heart.

There will always be a new generation of industry people. Why? Because individuals are drawn to it. There's no other industry like it. There's nothing like it in real life—sort of like the movies themselves.

The first movie of this volume is *The Extra Girl* and the last movie discussed is *The Errand Boy*. What an appropriate way to begin and end this book that celebrates creative careers in Hollywood. Don't give up the dream. There's a creative career for you. Here's to all of you, you future extra girls and errand boys . . . may you have much success.

CREATIVE CAREERS IN HOLLYWOOD FILMS

Chapter 1. ACTOR

- The Extra Girl (Associated Exhibitors, 1923)
- Show People (MGM, 1928)
- The Wild Party (American International, 1975)
- What Price Hollywood? (RKO-Pathé, 1932)
- A Star Is Born (United Artists, 1937)
- Going Hollywood (MGM, 1933)
- Showgirl in Hollywood (First National, 1930)
- Make Me a Star (Paramount, 1932)
- Merton of the Movies (MGM, 1947)
- Inside Daisy Clover (Warner Bros., 1965)
- The Purple Rose of Cairo (Orion, 1985)
- The Cowboy Star (Columbia, 1936)
- Won Ton Ton, the Dog that Saved Hollywood (Paramount, 1976)
- Under the Rainbow (Warner Bros., 1981)
- Sunset (TriStar, 1988)
- Dancing in The Dark (Twentieth Century Fox, 1949)
- Sunset Boulevard (Paramount, 1950)
- Dreamboat (Twentieth Century Fox, 1952)
- All About Eve (Twentieth Century Fox, 1950)
- The Star (Twentieth Century Fox, 1952)
- The Goddess (Columbia, 1958)
- A Star Is Born (Warner Bros., 1954)
- What Ever Happened to Baby Jane? (Warner Bros., 1962)
- The Patsy (Paramount, 1968)
- Hollywood Boulevard (New World, 1976)
- The Comic (Columbia, 1969)
- The Oscar (Embassy, 1966)
- Valley of The Dolls (Twentieth Century Fox, 1967)

- The Pickle (Columbia, 1993)
- Burn, Hollywood, Burn (Hollywood Pictures, 1997)
- Gods and Monsters (Lions Gate Films, 1998)
- The Truman Show (Paramount, 1998)
- American Movie (Sony Pictures Classics, 1999)

Chapter 6. PRESS

- Hollywood Speaks (Columbia, 1932)
- Bombshell (MGM, 1933)
- Hollywood Hotel (Warner Bros., 1938)
- Affairs of Annabel (RKO, 1938)
- Beloved Infidel (Twentieth Century Fox, 1959)
- The Big Knife (United Artists, 1955)

Chapter 7. PRODUCER

- Once In a Lifetime (Universal, 1932)
- Stand-in (United Artists, 1937)
- The Producers (MGM/UA, 1968)
- S.O.B. (Lorimar, 1981)
- . . . And God Spoke (Live Entertainment, 1993)
- Get Shorty! (MGM, 1995)
- Bowfinger (Universal, 1999)
- Matinee (Universal, 1993)

Chapter 8. PRODUCTION AND CREW

- Good Morning, Babylon (Vestron, 1987)
- Chaplin (TriStar, 1992)
- Day of The Locust (Paramount, 1975)
- Living in Oblivion (Sony Pictures Classics, 1995)
- The Stunt Man (Twentieth Century Fox, 1980)
- Hollywood Thrillmakers (Lippert, 1954)
- The Last Movie (Universal, 1971)
- The Great Waldo Pepper (Universal, 1975)

Chapter 9. STUDIO EXECUTIVE

- The Last Tycoon (Paramount, 1976)
- Man Of a Thousand Faces (Universal, 1957)
- The Bad and The Beautiful (MGM, 1952)
- The Player (Fine Line, 1992)
- Swimming with Sharks (TriMark, 1994)

Chapter 10. WRITER

- Boy Meets Girl (Warner Bros., 1938)
- Hearts of the West (MGM/UA, 1975)
- Barton Fink (Twentieth Century Fox, 1991)
- Without Reservations (RKO, 1946)
- The Way We Were (Columbia, 1973)
- In a Lonely Place (Columbia, 1950)
- Sunset Boulevard (Paramount, 1950)
- Susan Slept Here (RKO, 1954)
- Paris When It Sizzles (Paramount, 1964)
- The Big Picture (Columbia, 1989)
- The Player (Fine Line, 1992)
- My Life's in Turnaround (Islet, 1994)
- Get Bruce! (Miramax, 1999)
- Best Friends (Warner Bros., 1982)
- The Lonely Lady (Universal, 1983)
- The Muse (October Films, 1999)

Chapter 11. LOT LIFE

- Free & Easy (MGM, 1930)
- Movie Crazy (Paramount, 1932)
- Singin' in the Rain (MGM/UA, 1952)
- It's a Great Feeling (Warner Bros., 1949)
- Who Framed Roger Rabbit (Amblin/Touchstone, 1988)
- Falcon In Hollywood (RKO, 1944)
- Four Girls In Town (Universal, 1956)
- The Errand Boy (Paramount, 1961)

INDEX

ABOUT THE AUTHOR

Photo by Steven Baker

Struck by Daisy Clover's glamorous life as a Hollywood teenage starlet, Laurie Scheer, age 7, was determined to get into show business. Watching movies such as *Inside Daisy Clover, Sunset Boulevard,* and *The Errand Boy* only fueled her fire to work in the industry. Flash forward a few years, and, with her Broadcasting degree in hand, she found herself in Los Angeles to begin her own media career. Every part of the journey, from assistant to network president, has been rewarding. Laurie thinks everyone should follow their dreams, especially if their dreams originate from the movies.

When she's not writing, teaching, or producing, Laurie's cycling, visiting the beach at low tide, or studying astrology and Taoism.

BOOKS FROM ALLWORTH PRESS

An Actor's Guide — Your First Year in Hollywood, Revised Edition
by Michael Saint Nicholas (paperback, 6 × 9, 272 pages, $18.95)

Surviving Hollywood
by Jerry Rannow (paperback, 6 × 9, 224 pages, $16.95)

Acting for Film
by Cathy Haase (paperback, 6 × 9, 240 pages, $19.95)

Hollywood Dealmaking: Negotiating Talent Agreements
by Dina Appleton and Daniel Yankelevitz (paperback, 6 × 9, 256 pages, $19.95)

Producing for Hollywood: A Guide for Independent Producers
by Paul Mason and Don Gold (paperback, 6 × 9, 272 pages, $19.95)

Directing for Film and Television, Revised Edition
by Christopher Lukas (paperback, 6 × 9, 256 pages, $19.95)

Shoot Me: Independent Filmmaking from Creative Concept to Rousing Release
by Roy Frumkes and Rocco Simonelli (paperback, 6 × 9, 240 pages, $19.95)

The Health & Safety Guide for Film, TV & Theater
by Monona Rossol (paperback, 6 × 9, 256 pages, $19.95)

Technical Film and TV for Nontechnical People
by Drew Campbell (paperback, 6 × 9, 256 pages, $19.95)

The Filmmaker's Guide to Production Design
by Vincent LoBrutto (paperback, 6 × 9, 216 pages, $19.95)

The Screenwriter's Legal Guide, second edition
by Stephen F. Breimer (paperback, 6 × 9, 320 pages, $19.95)

Writing Television Comedy
by Jerry Rannow (paperback, 6 × 9, 224 pages, $14.95)

Career Solutions for Creative People
by Dr. Rhonda Ormont (paperback, 6 × 9, 320 pages, $19.95)

Get the Picture? The Movie Lover's Guide to Watching Films
by Jim Piper (paperback, 6 × 9, 240 pages, $18.95)

The Directors: Take One
by Robert J. Emery (paperback, 6 × 9, 416 pages, $19.95)

The Directors: Take Two
by Robert J. Emery (paperback, 6 × 9, 384 pages, $19.95)

Please write to request our free catalog. To order by credit card, call 1-800-491-2808 or send a check or money-order to Allworth Press, 10 East 23rd Street, Suite 210, New York, NY 10010. Include $5 for shipping and handling for the first book ordered and $1 for each additional book. Ten dollars plus $1 for each additional book if ordering from Canada. New York State residents must add sales tax.

To see our complete catalog on the World Wide Web, or to order online, you can find us at *www.allworth.com.*